Lecture Notes on Mathematical Modelling in the Life Sciences

Series Editors
Angela Stevens
Michael C. Mackey

For further volumes:
http://www.springer.com/series/10049

Moisés Santillán

Chemical Kinetics, Stochastic Processes, and Irreversible Thermodynamics

Springer

Moisés Santillán
Centro de Investigación y de Estudios
Avanzados del IPN (Cinvestav)
Unidad Monterrey
Vía del Conocimiento 201, Parque PIIT
Apodaca NL, México

ISSN 2193-4789 ISSN 2193-4797 (electronic)
ISBN 978-3-319-06688-2 ISBN 978-3-319-06689-9 (eBook)
DOI 10.1007/978-3-319-06689-9
Springer Cham Heidelberg New York Dordrecht London

Library of Congress Control Number: 2014941453

Mathematics Subject Classification (2010): 92C40, 92C45, 80A30, 82C31, 60H35, 34F05

Printed on acid-free paper

Springer is part of Springer Science+Business Media (www.springer.com)

To Matilda, our most recent family member, and the source of all inspiration!

To Mariana, my wife, whose love, support, and comprehension made this book possible!

To my parents, who have paved my path through life with their boundless love and wisdom!

To Pedro Santillán, Lourdes A. Vega-Acosta, Fernando Angulo-Brown, and Michael C. Mackey, my mentors!

Preface

There is a long tradition regarding mathematical modeling of biological phenomena, which dates back to the very beginning of modern science. However, there has been a revival in the last decade and a half, in which much effort has been made to elucidate the dynamic behavior of intracellular processes. This has been due in part to the development of impressive novel experimental techniques that permit precise measurements at the single cell, and even at the single molecule levels. But also thanks to the availability of low-cost and highly efficient computational power, which permits to numerically explore the dynamics of quite complex systems. One particular problem that has generated big interest is the stochasticity of gene expression.

Over the last years it has become clear that the dynamics of most biological phenomena can be studied via the techniques of either nonlinear dynamics or stochastic processes. In either case, the biological system is usually visualized as a set of interdependent chemical reactions and the model equations are derived out of this picture. Deterministic, nonlinear dynamic models rely on chemical kinetics, while stochastic models are developed from the chemical master equation. Recent publications have demonstrated that deterministic models are nothing but an average description of the behavior of unicellular stochastic models. In that sense, the most detailed modeling approach is that of stochastic processes. However, both the deterministic and the stochastic approaches are complementary. The vast amount of available techniques to analytically explore the behavior of deterministic, nonlinear dynamical models is almost completely inexistent for their stochastic counterparts. On the other hand, the only way to investigate biochemical noise is via stochastic processes.

It is my impression that the excitement of developing a new science (that has been termed systems biology) has made people unaware of a large amount of related discoveries performed in the late nineteenth and early twentieth centuries by people like Planck, Nernst, de Groot, Prigogine, etc., which we have classified as physical chemistry and irreversible thermodynamics. Moreover, although not as extensive a dismissal, there is no general acknowledgment on the part of systems biologists of the common mathematical grounds shared by biochemical and electrophysiological models; the latter ones are usually classified in the realm of biophysics. One of the objectives of the present book is to show that the approaches of deterministic nonlinear dynamics, stochastic chemical master equation, and irreversible thermodynamical chemistry are all complementary, and that their proper combination allows a deeper understanding of the dynamic behavior of a large variety of biological systems. In particular, we tackle in this book gene expression and ion transport across membranes.

There are many great books on nonlinear dynamics, stochastic processes, irreversible thermodynamics, physical chemistry, and biochemistry. Nonetheless, to the best of my knowledge, there is none that brings all of these theories together, in an introductory but formal and comprehensive manner, for people interested in modeling biological phenomena. The present book is aimed at filling, at least in part, this gap. In order to achieve this goal I decided to follow a hands-on constructivist approach. The theory is developed stepwise, starting from the simplest concepts, and building upon them to derive, one step at a time, a more general framework. But instead of first developing the theory and later studying its possible applications to biological systems, the examples are introduced right away. On the one hand, each theoretical development is motivated by specific biological examples. But also, every new mathematical derivation is immediately applied to one or more biological systems.

The target audience of this book are mainly last year undergraduate and graduate students with a solid mathematical background (physicists, mathematicians, and engineers), as well as with basic notions of biochemistry and cellular biology, who are interested in learning the previously described techniques to model biological phenomena. The book can also be useful to students with a biological background, who are interested in mathematical modeling, and have a working knowledge of calculus, differential equations, and basic notions of probability theory.

The book is organized as follows. Chapters 1 and 2 are introductory and in them some basic notions of chemical kinetics and thermodynamics are, respectively, presented. The readers already familiar with this material can jump directly to Chap. 3. However, I encourage everyone to at least take a look at Chaps. 1 and 2, because the material there introduced is widely employed in the rest of the book. In Chap. 3, the so-called telegraph stochastic process is analyzed from all the three perspectives discussed above, and the results are employed to discuss some aspects of ion channel gating, promoter repression and activation, and protein phosphorylation and dephosphorylation. In Chap. 4 the following stochastic processes are introduced and analyzed in connection with the production and degradation of biological molecules: Poisson process, exponential decay, and birth–death process. In Chap. 5, birth–death processes are generalized to account for enzyme kinetics. The concept of quasi-stationary approximation for stochastic processes is also introduced in Chap. 5. Chapter 6 is advocated to studying a generalization of the telegraph process in the context of the chemical interaction between one receptor and several ligands. Chapter 7 further generalized the results in Chap. 6 to account for cooperativity. In Chap. 8, all the results and developments introduced in the previous chapters are applied to the study of gene expression. Finally, in Chap. 9, the developed theory is applied to studying ion transport across membranes.

Like all long-term processes, the writing of the present book involved numerous people and institutions. In particular, I am in debt with my working place, the *Centro de Investigación y de Estudios Avanzados del Instituto Politécnico Nacional*, for granting me a sabbatical leave in which I found the necessary time to write. I also thank the Department of Physiology of McGill University and the International Centre for Theoretical Physics for their hospitality and for providing me with the most adequate environment to carry out this project. Finally, of all the people who were involved, one way or another, in the process of writing the present book, I wish to emphasize Michael C. Mackey, Emanuel Salazar-Cabazos, Román U. Zapuién-Campos, and Luis U. Aguilera-de-Lira. I am deeply grateful with all the persons who contributed to the present book, but the support and advice of the above-mentioned people were so important that, in strict sense, made this book possible. In spite of several reviews it is possible that some mistakes are found along the book. I, and no one else, am the only responsible for all of them.

Apodaca, Mexico Moisés Santillán

Contents

Chapter 1
Brief Introduction to Chemical Kinetics

Abstract In this chapter we present a brief introduction to chemical kinetics. Key concepts like: reversibility of chemical reactions, reaction rate, reaction rate constant, and chemical equilibrium, are introduced and discussed. The most important of the results here derived is the so-called law of mass action; which we discuss from the perspective of chemical kinetics. In this chapter we follow a heuristic rather than a formal approach. We start by analyzing a few simple chemical reactions to gain insight into the chemical kinetics basic concepts. After that, we heuristically derive and discuss the corresponding results for the most general case. The interested reader can consult any of the many available books on the subject. We particularly recommend the book by Houston (Chemical kinetics and reaction dynamics. McGraw-Hill, New York, 2001).

1.1 The Nature of Chemical Reactions

From a macroscopic perspective, a chemical reaction consists of a vast amount of sequential, individual, chemical steps. Each step takes place when one or more molecules of one or more chemical species (the reactants) interact via collisions and transform their chemical nature to give rise to a different set of molecules of distinct chemical species (the products). Strictly speaking, all chemical reactions are reversible because it is possible that the product molecules collide in such a way that they react and give rise to the reactant molecules. The usual way to represent these processes is as follows:

$$\alpha_1 A_1 + \alpha_2 A_2 + \cdots \rightleftharpoons \beta_1 B_1 + \beta_2 B_2 \cdots . \tag{1.1}$$

The above expression denotes a reaction in which the reactants are chemical species A_1, A_2, etc. while the products are chemical species B_1, B_2, etc. Furthermore, in each forward step, α_1 molecules of species A_1, α_2 molecules of species A_2, etc. react and disappear, giving rise to β_1 molecules of species B_1, β_2 molecules of

M. Santillán, *Chemical Kinetics, Stochastic Processes, and Irreversible Thermodynamics*, Lecture Notes on Mathematical Modelling in the Life Sciences,
DOI 10.1007/978-3-319-06689-9_1, © Springer International Publishing Switzerland 2014

species B_2, etc. As previously mentioned, this chemical reaction is reversible and so the individual steps can take place in both directions. The right-up and left-down harpoons denote the chemical reaction reversibility.

1.2 Reaction Rate: A Very Simple Example

Consider the following chemical reaction

$$A + B \rightleftharpoons C. \tag{1.2}$$

The reaction rate v is defined as the net number of forward steps (forward steps minus backward steps) taking place per unit time, per unit volume. In this particular example, v equals the rate per unit volume at which C molecules appear, or equivalently, the rate per unit volume at which A or B molecules disappear. If we take into consideration the fact that the reaction is reversible, one can define forward and backward reaction rates, v^+ and v^-, as respectively the number of forward and backward individual steps taking place per unit time, per unit volume. In terms of v^+ and v^-, the net reaction rate (also known as speed) is

$$v = v^+ - v^-.$$

Consider for now the forward reaction. In order for one A molecule and one B molecule to react, they need to collide. Thus, the forward reaction rate is proportional to the number of such collisions. All the molecules in a chemical reactor follow Brownian random trajectories due to thermal agitation. In particular, the higher the temperature, the larger the average velocity of all molecules, and so the higher the collision probability and the reaction rate. The chemical reaction rate is also affected by other factors. For instance, as the number of A molecules increases, the probability that one of them collides with a B molecule augments. In fact, it can be shown that the forward reaction velocity is proportional to n_A: the A molecule count. Using the same reasoning, we can argue that the forward reaction rate is also proportional to the B molecule count: n_B. Finally, the reactor volume, V, also plays an important role. In a larger volume it is less probable that any of the existing B molecules collide with a given A molecule; and vice versa. By means of collision theory (Houston 2001) it is possible to demonstrate that

$$v^+ = k^+ \frac{n_A}{V} \frac{n_B}{V}, \tag{1.3}$$

where k^+ is a function of the temperature. However, under isothermal conditions, it can be regarded as a constant. Constant k^+ is called forward-reaction rate constant.

Let us assume that the volume of the reservoir in which the reaction takes place (the reactor) is constant, and define the forward-reaction molecular velocity (or simply the velocity), $\mathscr{V}^+$, as the net number of forward individual steps taking place per unit time: $\mathscr{V}^+ = v^+ V$. With this, Eq. (1.3) can be rewritten as

$$\mathscr{V}^+ = \kappa^+ n_A n_B, \tag{1.4}$$

with $\kappa^+ = k^+/V$.

Molecules C can be destabilized by collisions with any other molecule, and split into molecules A and B. Therefore, one can argue that the backward reaction rate is given by

$$v^- = k^- \frac{n_C}{V}, \tag{1.5}$$

with k^- the backward-reaction rate constant. From Eq. (1.5), the corresponding backward molecular velocity is

$$\mathscr{V}^- = \kappa^- n_C, \tag{1.6}$$

In which $\kappa^- = k^-/V$.

From Eqs. (1.3) and (1.5), the global reaction rate is

$$v = v^+ - v^- = k^+ c_A c_B - k^- c_C, \tag{1.7}$$

Where $c_A = n_A/V$, $c_B = n_B/V$, and $c_C = n_C/V$ are the concentrations of molecules A, B, and C, respectively. Similarly, the global molecular velocity can be written from (1.4) and (1.6) as

$$\mathscr{V} = \mathscr{V}^+ - \mathscr{V}^- = \kappa^+ n_a n_B - \kappa^- n_C. \tag{1.8}$$

1.3 Dynamic Equations

Let us regard for a moment the general chemical equation in Eq. (1.1), and assume that it takes place at rate v. Recall that the reaction rate is the net number of forward individual steps taking place per unit time, per unit volume. Then, if we consider that each time a forward chemical step takes place α_i molecules of species A_i disappear, while β_i molecules of species B_i appear, then the rate of change for the concentrations of all the chemical species involved in the reaction in (1.1) is as follows:

$$\frac{dc_{A_i}}{dt} = -\alpha_i v, \quad \frac{dc_{B_i}}{dt} = -\beta_i v. \tag{1.9}$$

Thus, in the case of reaction (1.2), the rates of change for the concentrations of all the involved chemical species are—see Eq. (1.7):

$$-\frac{dc_A}{dt} = -\frac{dc_B}{dt} = \frac{dc_C}{dt} = v = k^+ c_A c_B - k^- c_C. \tag{1.10}$$

We can also write an equivalent equation for the rates of change of the molecular counts as follows

$$-\frac{dn_A}{dt} = -\frac{dn_B}{dt} = \frac{dn_C}{dt} = \mathcal{V} = \kappa^+ n_A n_B - \kappa^- n_C. \tag{1.11}$$

1.4 Chemical Equilibrium

Chemical equilibrium is a key concept in chemical kinetics. It is achieved when the concentrations or the molecular counts of all the species involved in the chemical reaction reach a stationary value. According to (1.9), the reaction in (1.2) reaches chemical equilibrium when

$$\frac{k^-}{k^+} = \frac{\overline{c}_A \overline{c}_B}{\overline{c}_C}, \tag{1.12}$$

where $\overline{c}_A$, $\overline{c}_B$, and $\overline{c}_C$ respectively denote the A, B, and C stationary concentrations. This equation does not suffice to determine the stationary concentration values (for that, more information is necessary). However, it establishes a fundamental relationship that has to be satisfied by the steady concentrations of all the reaction chemical species, independently of their initial values. This relationship is called the law of mass action.

The reader is invited to prove that, in terms of molecular counts, the law of mass action for the reaction in (1.2) takes the following form

$$\frac{\kappa^-}{\kappa^+} = \frac{\overline{n}_A \overline{n}_B}{\overline{n}_C}, \tag{1.13}$$

where $\overline{n}_A$, $\overline{n}_B$, and $\overline{n}_C$ respectively denote the A, B, and C stationary molecule counts.

1.5 Second Order Chemical Kinetics

The kinetics of the reaction in Eq. (1.2) are first order in all the chemical species' concentrations. The reason for this is that all the concentrations appear to the power one in the forward and backward reaction rates. The reader would have already guessed that not all reactions are like that. For instance, consider the following example:

$$2A \rightleftharpoons C. \tag{1.14}$$

To derive the expressions for the corresponding forward and backward reaction rates simply consider that chemical species A takes the place of B in the reaction in (1.2). From this, it is straightforward to obtain the following:

$$v^+ = k^+ c_A^2, \quad v^- = k^- c_C. \tag{1.15}$$

Notice that the exponent of c_A and c_C are nothing but the stoichiometric coefficients of the corresponding chemical species in (1.2).

By following the reasoning leading to Eq. (1.9), the differential equations governing the dynamical evolution of c_A and c_C result to be

$$-\frac{1}{2}\frac{dc_A}{dt} = \frac{dC_c}{dt} = v^+ - v^- = k^+ c_A^2 - k^- c_C. \tag{1.16}$$

Furthermore, the law of mass action in this case takes the following form:

$$\frac{k^-}{k^+} = \frac{c_a^2}{c_c}. \tag{1.17}$$

By comparing Eqs. (1.17) and (1.12) we note that the ratio of the backward to the forward reaction-rate constants appears on the left-hand side of both equations. Moreover, on the right-hand side we find a fraction whose numerator has the reactant concentrations to the corresponding stoichiometric coefficient, and whose denominator has the concentration of the reaction product. In the following section we shall see that, indeed, the law of mass action has this form in general.

All the results in this section can also be written in terms of molecule counts, rather than concentrations. The readers are invited to do so.

1.6 The General Case

Let us now analyze the most general chemical reaction presented in Eq. (1.1). We can generalize the particular cases discussed in the previous sections as follows. The forward and backward reaction rates for the reaction in (1.1) are

$$v^+ = k^+ c_{A_1}^{\alpha_1} c_{A_2}^{\alpha_2} \cdots, \quad v^- = k^- c_{B_1}^{\beta_1} c_{B_2}^{\beta_2} \cdots. \tag{1.18}$$

In the equation above k^+ and k^+ denote the corresponding rate constants. Notice that Eqs. (1.3), (1.5), and (1.15) are particular cases of Eq. (1.18). Moreover, the differential equations governing the dynamics of c_{A_i} and c_{B_i} are:

$$-\frac{1}{\alpha_i}\frac{dc_{A_i}}{dt} = \frac{1}{\beta_i}\frac{dc_{B_i}}{dt} = v^+ - v^- = k^+ \prod_i c_{A_i}^{\alpha_i} - k^- \prod_i c_{B_i}^{\beta_i}. \tag{1.19}$$

From this, the condition for chemical equilibrium results to be

$$K_D = \frac{k^-}{k^+} = \frac{\prod_i c_{A_i}^{\alpha_i}}{\prod_i c_{B_i}^{\beta_i}}. \tag{1.20}$$

This is the most general form of the law of mass action. The fraction k^-/k^+ is known as the reaction dissociation constant. In the following chapter we shall see that the law of mass action not only has a kinetic significance, as we have just studied, but also a thermodynamic one. However, the reader will have to keep on reading the book in order to gain a deeper understanding about this connection.

1.7 Summary

This chapter is meant as a brief introduction to chemical kinetics. Some central concepts, like reaction rate and chemical equilibrium, have been introduced and their meaning has been reviewed. We have further seen how to employ those concepts to write a system of ordinary differential equations to model the time evolution of the concentrations of all the chemical species in the system. The resulting equations can then be numerically or analytically solved, or studied by means of the techniques of nonlinear dynamics. A particularly interesting result obtained in this chapter was the law of mass action, which dictates a condition to be satisfied for the equilibrium concentrations of all the chemical species involved in a reaction, regardless of their initial values. In the forthcoming chapters we shall use this result to connect different approaches like chemical kinetics, thermodynamics, etc.

Chapter 2
Brief Introduction to Thermodynamics

Abstract This chapter is devoted to introducing the basic concepts of thermodynamics, specially as applied to chemistry. The reader must be aware that the material in this chapter is rather technical and succinct. Therefore, it is quite possible that some of the results are not clear, even after carefully reading the chapter material more than once. Of course, people interested in this field can go to the specialized literature. However, one of the mayor goals of the present book is to help make these things clear through some examples. So, if things seem a bit blurry after finishing this chapter, please do not despair and keep reading. It will soon get better, promise.

2.1 The First and Second Laws of Thermodynamics

Let us start by briefly reviewing some of the most important concepts in thermodynamics, beginning with the first law. The readers interested in reading about this subject with more detail are referred to the following books: (Planck 1945; Ben-Naim 2007; Beard and Qian 2008).

Denote by E the energy of the system under study. According to the first law of thermodynamics, E can change because energy in the form of heat enters the system, because mechanical work is performed on the system, or because the molecular counts of the various chemical species composing the system change (chemical work). In particular, if the system is a compressible fluid, the first law of thermodynamics can be written as (Planck 1945):

$$dE = đQ - PdV + \sum_i \mu_i dN_i. \tag{2.1}$$

M. Santillán, *Chemical Kinetics, Stochastic Processes, and Irreversible Thermodynamics*, Lecture Notes on Mathematical Modelling in the Life Sciences,
DOI 10.1007/978-3-319-06689-9_2, © Springer International Publishing Switzerland 2014

In the above equation $đQ$ denotes the amount of heat entering the system (the symbol $đ$ represents an inexact differential), $-PdV$ is the amount of mechanical work performed on the system (P is the hydrostatic pressure and V is the system volume), and μdN is the so-called chemical work performed on the system (μ_i and N_i are respectively the chemical potential and the molecular count of the ith chemical species).

The second law of thermodynamics introduces a new variable called the entropy, usually denoted by S, which satisfies the following relation (Planck 1945):

$$TdS \geq đQ. \tag{2.2}$$

That is, the entropy increment of a system along a given process is always larger, or at least equal, than the influx of heat along such process, divided by the temperature T. The equality is satisfied when the system undergoes a reversible process.

In the particular case in which $đQ = 0$ (i.e. when the system suffers an adiabatic process) the second law takes the form:

$$dS \geq 0.$$

Meaning that the entropy of an isolated system never decreases. If we further consider that any isolated system evolves to a thermodynamic equilibrium state through an irreversible process, the above result implies that the equilibrium state is characterized by having the maximum possible entropy compatible with the constrains of constant internal energy, volume, and particle counts. This last result is known as the maximum entropy principle.

It is also possible to express the second law of thermodynamics as optimality principles in some other particular cases. For instance, by combining Eqs. (2.1) and (2.2) we obtain

$$dE \leq TdS - PdV + \sum_i \mu_i dN_i. \tag{2.3}$$

Hence, if S, V, and N_i are kept constant, then

$$dE \leq 0.$$

This means that the energy of a system kept at constant entropy, volume, and particle count can only decrease. Or equivalently, that the equilibrium state of a system subject to these constrains possesses the minimum possible energy compatible with them.

Yet another instance of second law of thermodynamics can be derived by introducing a new thermodynamic quantity, called the Gibbs free energy (Planck 1945):

$$G = E - TS - PV. \tag{2.4}$$

By differentiating Eq. (2.4) and substituting Eq. (2.3) we get

$$dG \leq -Sdt + VdP + \sum_i \mu_i dN_i. \tag{2.5}$$

Hence, the Gibbs free energy of a system that is kept at constant T, P, and N_i can only decrease, and so the corresponding equilibrium state is characterized by having the minimum possible G value.

Under the assumption that the system undergoes a reversible process Eq. (2.5) becomes

$$dG = SdT + VdP + \sum_i \mu_i dN_i.$$

Furthermore, since most biochemical processes take place at constant pressure and temperature, the last expression reduces under such conditions to

$$dG = \sum_i \mu_i dN_i. \tag{2.6}$$

Keep this last result in mind because we will extensively use it in the rest of the book.

2.2 Thermodynamics of Chemical Reactions

Assume that the molecular counts of all the chemical species change because of chemical reactions taking place within the system. Thus (de Groot and Mazur 2013):

$$dN_i = \sum_\lambda \delta_{i\lambda} d\zeta_\lambda,$$

in which the sum is carried out over all the chemical reactions, $\delta_{i\lambda}$ is an stoichiometric coefficient that gives the change in the number of molecules of chemical species i when an individual event of the λth reaction occurs, and ζ_λ is the degree of advance of the λth reaction (it measures the number of individual events that have taken place since the beginning of the experiment). If the reactions are coupled, then all of them advance at the same rate ($\zeta_\lambda = \zeta$ for all λ), and

$$dN_i = \sum_\lambda \delta_{i\lambda} d\zeta, \tag{2.7}$$

Substitution of (2.7) into (2.6) leads to

$$\Delta G = \frac{dG}{d\zeta} = \sum_\lambda \sum_i \delta_{i\lambda} \mu_i. \tag{2.8}$$

The newly defined quantity ΔG is the so-called total free energy change of the system due to the undergoing chemical reactions. A free energy change can be defined for each reaction:

$$\Delta G_\lambda = \sum_i \delta_{i\lambda} \mu_i.$$

And so

$$\Delta G = \sum_\lambda \Delta G_\lambda.$$

For simplicity, consider in what follows a chemical system in which only one chemical reaction is taking place. Stoichiometric coefficients can be positive (if the corresponding chemical species is a product of the reaction) or negative (if the chemical species is a reactant). Let us denote by $-\alpha_i$ all the negative stoichiometric coefficients, and by β_i all the positive ones. With this, Eq. (2.8) can be rewritten as

$$\Delta G = \sum_i \beta_i \mu_i - \sum_j \alpha_j \mu_j. \tag{2.9}$$

In the above equation we have omitted the sum over λ because of the assumption that only one chemical reaction occurs. Moreover, subindexes i and j respectively denote the reaction products and reactants.

In the following section we derive an expression for the chemical potential in terms of the system state variables. But for now let us just take the final result:

$$\mu = \mu^O + k_B T \ln \frac{c}{c^O}, \tag{2.10}$$

where $c = N/V$ is the molecule concentration of the corresponding chemical species (N is the molecule count and V is the volume), c^O and μ^O are respectively the molecule concentration and the chemical potential under reference conditions, and k_B is Boltzmann's constant. Thus, if we assume without loss of generality that the reference conditions are chosen in such a way that $c^O V = 1$, Eq. (2.10) transforms into

$$\mu = \mu^O + k_B T \ln N. \tag{2.11}$$

Substitution of Eq. (2.11) into (2.9) leads to

$$\Delta G = k_B T \sum_i \beta_i \ln \left(N_i e^{\mu_i^O / k_B T} \right) - k_B T \sum_j \alpha_j \ln \left(N_j e^{\mu_j^O / k_B T} \right).$$

We have seen that thermodynamic equilibrium implies that $dG = 0$. Then, it follows from (2.8) that a chemical reaction reaches equilibrium when $\Delta G = 0$, and so when

$$\frac{\prod_i \overline{N}_i^{\alpha_i}}{\prod_j \overline{N}_j^{\beta_i}} = \frac{e^{\sum_i \beta_i \mu_i^O / k_B T}}{e^{\sum_j \alpha_j \mu_j^O / k_B T}}, \qquad (2.12)$$

where $\overline{N}_i$ denotes the equilibrium molecule count of the corresponding chemical species. Observe that this equation is strikingly similar to Eq. (1.20), suggesting that

$$K_D = \frac{e^{\sum_i \beta_i \mu_i^O / k_B T}}{e^{\sum_j \alpha_j \mu_j^O / k_B T}}. \qquad (2.13)$$

In the following chapters we shall prove that Eqs. (2.12) and (1.20) are identical and indeed Eq. (2.13) is true.

2.3 Understanding the Chemical Potential Concept

Consider a system that has a countable number of available states and suppose that we know the probability (p_i) associated with each one of them. The system entropy can then be computed as (Ben-Naim 2007):

$$S = -k_B \sum_i p_i \ln p_i, \qquad (2.14)$$

The problem with this approach is that one usually does not know the probability distribution p_i, but only the constraints the system is subject to. In order to solve this dilemma, one can make use of the second law of thermodynamics in one of its following versions (Planck 1945):

- Under conditions of constant energy (E), volume (V), and number of particles (N), the equilibrium state maximizes the system entropy (S).
- Under conditions of constant temperature (T), volume, and number of particles, the equilibrium state minimizes the Helmholtz free energy defined as: $F = E - TS$.
- Under conditions of constant temperature (T), pressure (P), and number of particles, the equilibrium state minimizes the Gibbs free energy defined as: $G = E - TS - PV$.

Let us focus in the second case. The system average energy can be computed in terms of the probability distribution as:

$$E = \sum_i \epsilon_i p_i, \tag{2.15}$$

in which ϵ_i is the energy associated with the ith state. Then, by combining Eqs. (2.14) and (2.15) we obtain the following expression for the Helmholtz free energy in terms of the probability distribution:

$$F = \sum_i p_i(\epsilon_i + k_B T \ln p_i). \tag{2.16}$$

With this, we can figure out what the equilibrium probability distribution is when the system temperature, volume, and number of particles are constrained. All we need to do is to find the probability distribution that minimizes Eq. (2.16), subject to the normalization condition:

$$\sum_i p_i = 1.$$

This a classical problem of calculus that can be solved using the technique of Lagrange multipliers. We leave for the reader to prove that the solution is the renowned Boltzmann distribution:

$$P_i^{eq} = \frac{e^{-\epsilon_i/k_B T}}{Z}, \tag{2.17}$$

where

$$Z = \sum_i e^{-\epsilon_i/k_B T} \tag{2.18}$$

is known as the partition function. Substitution of Eq. (2.17) into Eq. (2.16) finally leads to the following expression for F:

$$F = -k_B T \ln Z. \tag{2.19}$$

A molecule in solution can be seen as a system complying with constant T, V, and N. Let $Z^{(1)}$ denote its partition function. Having the partition function a probabilistic interpretation, the partition function of a composite system equals the product of the components' partition functions, unless they are identical and indistinguishable. Consider for instance a system composed of N identical, indistinguishable molecules in solution. In this case, the system partition function is:

$$Z = \frac{{Z^{(1)}}^N}{N!}. \tag{2.20}$$

$N!$ is the number of ways in which the N molecules can be permuted. By introducing the factor $N!$ we have taken into account the fact that, due to the indistinguishability of molecules, all of the permutations account for the same state. In other words, by dividing by $N!$ one avoids to count the same state multiple times while computing the system free energy—see Eq. (2.19).

By substituting Eq. (2.20) into Eq. (2.19) we obtain for Helmholtz free energy the following expression:

$$F = k_B T (N \ln N - N - N \ln Z^{(1)}). \tag{2.21}$$

In the derivation of Eq. (2.21) we have made use of Stirling's approximation: $\ln N! \approx N \ln N - N$.

If we differentiate $F = E - TS$ and take into consideration Gibbs fundamental relation ($dE = TdS - PdV + \sum_i \mu_i dN_i$) we get

$$dF = SdT - PdV + \mu dN.$$

Hence

$$\mu = \left(\frac{\partial F}{\partial N} \right)_{T,V}. \tag{2.22}$$

Finally, substitution of (2.21) into (2.22) leads to

$$\mu = k_B T \ln N - k_B T \ln Z^{(1)}.$$

By introducing a few extra factors, without altering the equality, this equation can be rewritten as

$$\mu = \mu^O + k_B T \ln \frac{c}{c^O}, \tag{2.23}$$

in which $c = N/V$ is the molecule concentration, c^O is the concentration at some standard conditions, and

$$\mu^O = -k_B T \ln \frac{Z^{(1)}}{Vc^O}$$

is the system chemical potential when $c = c^O$, which depends only on the molecules' chemical nature. Equation (2.23) is the standard formula for the chemical potential found in textbooks (Beard and Qian 2008).

2.4 Summary

This chapter was devoted to introducing the thermodynamic concepts and formalism essential to understand chemical reactions. Thus, we reviewed the first and second laws of thermodynamics, introduced the concept of thermodynamic equilibrium, defined the free energy change, and used it to prove that thermodynamic and chemical equilibrium are equivalent concepts. Interestingly, we were able to obtain the las of mass action from purely thermodynamic considerations, suggesting that the thermodynamic and the chemical kinetics approaches are closely related. This connection is explored in detail in the next chapter. Finally, the last section of the chapter was dedicated to understanding the concept of chemical potential from the perspective of statistical mechanics. In later chapters we tackle this same question from different angles.

Chapter 3
Different Approaches to Analyzing a Simple Chemical Reaction

Abstract In this chapter, the dynamics of the chemical reaction $A \rightleftharpoons B$ are analyzed from the perspectives of macroscopic chemical kinetics, thermodynamics, and stochastic processes. The main objective is to show that all these approaches are not exclusive but complementary, and that properly combining them through a fourth unifying approach (the one I call the energy landscape approach) allows a deeper understanding of the system dynamic behavior. After finishing this chapter not only the reader shall understand the different approaches to analyzing the dynamics of a chemical reaction, but will also obtain some basic notions of stochastic processes. Namely, the chemical master equation and Gillespie's algorithm.

3.1 The Chemical Kinetics Approach

Let us begin by introducing a simple chemical reaction (perhaps the simplest possible one):

$$A \rightleftharpoons B. \tag{3.1}$$

As we discussed in Chap. 1, this reaction (like all other ones) is reversible. Therefore (3.1) denotes two complementary chemical reactions. The first one, in which a molecule of the chemical species A turns into a molecule of the species B, is represented by the right harpoon. The second reaction, corresponding to a molecule of species B turning into a molecule of species A, is represented by the left harpoon. Despite its simplicity, this reversible reaction-set is actually a good model for some essential biochemical processes like: the gating of an ion channel between the close and open states, a protein flipping between two different conformational states, and the switching of a promoter between the active and repressed states.

Although the following assertions are valid for all the chemical reactions that can be depicted by Eq. (3.1), I believe it is easier if one has an specific example

M. Santillán, *Chemical Kinetics, Stochastic Processes, and Irreversible Thermodynamics*, Lecture Notes on Mathematical Modelling in the Life Sciences,
DOI 10.1007/978-3-319-06689-9_3,

in mind. Hence, without loss of generality, consider a molecule that switches back and forth between conformational states A and B. As discussed in Chap. 1, what makes a molecule switch between its two available states are the constant, numerous collisions with the molecules in the surrounding medium (a phenomenon known as thermal agitation or noise). Regard now several identical such molecules, all of them subject to the same environmental conditions. Then, from the results in Chap. 1, the forward and backward reaction rates are:

$$\upsilon_{AB} = \kappa_{AB} n_A, \qquad \upsilon_{BA} = \kappa_{BA} n_B. \tag{3.2}$$

In the above equation υ_{XY} denotes the rate of the X to Y reaction, n_X corresponds to the number of molecules in state X, and the proportionality constants κ_{XY} are called the reaction rate constants.

If we assume a constant total number of molecules

$$n_A(t) + n_B(t) = n_T, \tag{3.3}$$

it follows from Eq. (3.2) that the time evolution of n_A is governed by:

$$\frac{dn_A}{dt} = -\upsilon_{AB} + \upsilon_{BA} = \kappa_{BA} n_T - (\kappa_{AB} + \kappa_{BA}) n_A. \tag{3.4}$$

We leave for the reader to prove that the solution to Eq. (3.4) is:

$$n_A(t) = \frac{\kappa_{BA}}{\kappa_{AB} + \kappa_{BA}} n_T + \left(n_A^O - \frac{\kappa_{BA}}{\kappa_{AB} + \kappa_{BA}} n_T \right) e^{-(\kappa_{AB} + \kappa_{BA})t}, \tag{3.5}$$

in which n_A^O represents the initial count of A molecules. Equations (3.3) and (3.5) further imply that

$$n_B(t) = \frac{\kappa_{AB}}{\kappa_{AB} + \kappa_{BA}} n_T + \left(n_B^O - \frac{\kappa_{AB}}{\kappa_{AB} + \kappa_{BA}} n_T \right) e^{-(\kappa_{AB} + \kappa_{BA})t}, \tag{3.6}$$

where $n_B^O = n_T - n_A^O$ is the initial number of B molecules.

Observe from Eqs. (3.5) and (3.6) that both $n_A(t)$ and $n_B(t)$ respectively converge in an exponential fashion to

$$\overline{n}_A = \frac{k_{BA}}{k_{AB} + k_{BA}} n_T, \tag{3.7}$$

$$\overline{n}_B = \frac{k_{AB}}{k_{AB} + k_{BA}} n_T, \tag{3.8}$$

as $t \to \infty$. In other words, the dynamical system given by (3.3) has a unique, globally stable steady state given by Eq. (3.7)—recall that the dynamics of $n_B(t)$ are completely determined by those of $n_A(t)$. Finally, the speed of convergence is given by the sum of the reaction rate constants: the larger they are, the faster the systems converges to its stationary state.

We have from Eqs. (3.7) and (3.8) that

$$\frac{\overline{n}_A}{\overline{n}_B} = \frac{k_{BA}}{k_{AB}} = K_D. \tag{3.9}$$

The above equation is the law of mass action as applied to reaction (3.1). Recall that constant K_D is called the dissociation constant, while its inverse $K_A = K_D^{-1}$ is called the association constant (Houston 2001).

In conclusion, the chemical kinetics formalism is capable of predicting the time evolution of the number of molecules in each state. However, its validity rests upon the assumption that the number of molecules in states A and B is very large. In what follows we shall analyze the reasons behind this assertion.

3.2 Chemical Master Equation

Consider once more a single molecule flipping back and forth between states A and B. Since this is an stochastic process, it is impossible to predict in which state the molecule is going to be at any given time. Therefore, a probabilistic description is necessary. Let $P_A(t)$ be the probability that the molecule is in state A at time t. The probability that it is in state B at time t is then $P_B(t) = 1 - P_A(t)$. Assume that the probabilities per unit time that the molecules shifts from state A to sate B, and vice versa, are constant and respectively denoted by α_{AB} and α_{BA}. This means that, if the molecule is in state A, the probability that it shifts to state B in a time interval of length τ is $\alpha_{AB}\tau$. Similarly, the probability that the system flips from B to A in the same interval is $\alpha_{BA}\tau$. The probabilities per unit time α_{AB} and α_{BA} are usually called propensities (Gillespie 1977; Van Kampen 1992). Having the above discussion in mind, we can derive the equation governing the dynamics of $P_A(t)$ as follows. Assume that τ is short enough so that, at most, the system flips once from one state to the other. Then

$$P_A(t + \tau) = P_A(t)(1 - \alpha_{AB}\tau) + (1 - P_A(t))\alpha_{BA}\tau.$$

That is, we can find the system in state A at time $t + \tau$ if it was already there at time t and did not shift to state B during the period $[t, t + \tau]$, or if it was in state B at time t and flipped to state A during the same period. It follows after a little algebra that

$$\frac{P_A(t + \tau) - P_A(t)}{\tau} = -P_A(t)\alpha_{AB} + (1 - P_A(t))\alpha_{BA}.$$

Then, by taking the limit $\tau \to 0$ we obtain

$$\frac{dP_A(t)}{dt} = -P_A(t)\alpha_{AB} + (1 - P_A(t))\alpha_{BA}. \tag{3.10}$$

Equation (3.10) is called the master equation for $P_A(t)$ (Gillespie 1977; Van Kampen 1992). Notice that this is the same differential equation as that in (3.4). Hence, its solution is

$$P_A(t) = \frac{\alpha_{BA}}{\alpha_{AB} + \alpha_{BA}} + \left(P_A^O - \frac{\alpha_{BA}}{\alpha_{AB} + \alpha_{BA}}\right) e^{-(\alpha_{AB}+\alpha_{BA})t}, \tag{3.11}$$

in which P_A^O is the initial condition for $P_A(t)$. Finally, the fact that $P_A(t) + P_B(t) = 1$ further implies that

$$P_B(t) = \frac{\alpha_{AB}}{\alpha_{AB} + \alpha_{BA}} + \left(P_B^O - \frac{\alpha_{AB}}{\alpha_{AB} + \alpha_{BA}}\right) e^{-(\alpha_{AB}+\alpha_{BA})t}, \tag{3.12}$$

where $P_B^O = 1 - P_A^O$ is the initial condition for $P_B(t)$. Interestingly, we have from Eqs. (3.11) and (3.12) that $P_A(t)$ and $P_B(t)$ respectively converge to

$$\overline{P}_A = \frac{\alpha_{BA}}{\alpha_{AB} + \alpha_{BA}}, \tag{3.13}$$

$$\overline{P}_B = \frac{\alpha_{AB}}{\alpha_{AB} + \alpha_{BA}}. \tag{3.14}$$

This means that, regardless of the initial condition, the probabilities of finding the channel in its two available states reach constant values, fully determined by the propensities α_{AB} and α_{BA}.

The similarity between Eqs. (3.5)–(3.8) and Eqs. (3.11)–(3.14) suggests a relation between the propensities α_{AB} and α_{BA}, on the one hand, and the reaction rate constants κ_{AB} and κ_{BA}, on the other. This relation indeed exists and is not mere coincidence. In order to better understand it let us consider a system consisting of n_T independent, identical molecules like those just analyzed. Given their independence, the probability of finding n such molecules in state A (and the rest, $n_T - n$, in state B) obeys a binomial distribution (Evans et al. 2000):

$$\mathscr{P}_A(n,t) = \frac{n_T!}{n!(n_T - n)!} P_A(t)^n (1 - P_A(t))^{n_T - n}. \tag{3.15}$$

This result further implies that $\mathscr{P}_A(n,t)$ converges to a stationary probability distribution:

$$\overline{\mathscr{P}}_A(n) = \lim_{t\to\infty} \mathscr{P}_A(n,t) = \frac{n_T!}{n!(n_T - n)!} \overline{P}_A^n (1 - \overline{P}_A)^{n_T - n}. \tag{3.16}$$

Finally, the probability of having n molecules in state B at time can be straightforwardly calculated as $\mathscr{P}_B(n,t) = \mathscr{P}_A(n_T - n, t)$.

Another way to get the results in (3.15) and (3.16) is to write the master equation for $\mathscr{P}_A(n,t)$ and solve it. The reader is invited to demonstrate that such master equation is:

$$\begin{aligned}\frac{d\mathscr{P}_A(n,t)}{dt} &= \mathscr{P}_A(n-1,t)(n_T - n + 1)\alpha_{BA} \\ &-\mathscr{P}_A(n,t)(n_T - n)\alpha_{BA} \\ &+\mathscr{P}_A(n+1,t)(n+1)\alpha_{AB} \\ &-\mathscr{P}_A(n,t)n(\alpha_{BA} + \alpha_{AB}),\end{aligned} \tag{3.17}$$

and that the expressions in (3.15) and (3.16) are respectively its general and its stationary solutions.

With the probability distributions $\mathscr{P}_A(n,t)$ and $\mathscr{P}_B(n,t)$ it is possible to compute the average number of molecules in each state at all times, $N_A(t)$ and $N_B(t)$. Actually, it is not hard to prove from the properties of the binomial distribution (Evans et al. 2000) that

$$N_A(t) = n_T P_A(t), \tag{3.18}$$

$$N_B(t) = n_T P_B(t). \tag{3.19}$$

After substituting Eqs. (3.11) and (3.12) into Eqs. (3.18) and (3.19) we recover Eqs. (3.5) and (3.6), provided that $n_A^O = n_T P_A^O$, $n_B^O = n_T P_B^O$, $k_{AB} = \alpha_{AB}$, and $k_{BA} = \alpha_{BA}$. This confirms that, as we suspected, the reaction rate constants of the chemical-kinetics description correspond to the propensities of the chemical-master-equation description. Moreover, the differential equations of the model derived in terms of chemical kinetics are those governing the time evolution of the mean number of molecules in each state. But, when is this a good description? To answer this question we need to calculate the standard deviations corresponding to $N_A(t)$ and $N_B(t)$. Once more, it is straightforward to see, from the properties of the binomial distribution (Evans et al. 2000), that they are given by

$$\sigma_A(t) = \sigma_B(t) = \sqrt{n_T P_A(t) P_B(t)}. \tag{3.20}$$

The coefficient of variation, defined as the ratio of the standard deviation to the mean, measures how much the realizations of a random variable deviate from the mean value. It follows from Eqs. (3.18)–(3.20) that the coefficient of variation for the number of molecules in both states is inversely proportional to $\sqrt{n_T}$. Therefore, the larger the total number of molecules, the more accurate the description in terms of chemical kinetics. When the number of molecules is of the order of Avogadro's number, the coefficient of variation is negligible ($\sim 10^{-11}$). However, when the number of molecules is of the order of tens of thousands, the coefficient of variation is about $\sim 10^{-2}$; and when the molecular count is of the order of a few hundreds, the coefficient of variation is around $\sim 10^{-1}$. Thus, a description in terms of the chemical master equation becomes mandatory for low molecular counts. But still, the law-of-mass-action deterministic description is capable of providing some valuable information on the average values.

3.3 Gillespie Algorithm

Given the simplicity of the current system, it was possible to analytically solve the resulting chemical master equation. However, this is not always the case and one is limited to simulating individual realizations of the stochastic process in order to reconstruct the probability distributions out from several simulations. Below, we introduce the celebrated Gillespie algorithm (Gillespie 1977) to simulate the stochastic evolution of continuous-time discrete-state stochastic processes, like the one analyzed in the present chapter.

Let us start with a single molecule flipping back and forth between states A and B. Assume that it arrived in state A at time $t = 0$, and let us compute the probability that the molecule shifts back to state B in the interval $[\tau, \tau + dt]$. This probability can be written as:

$$p_A(\tau)dt = p_A^o(\tau)\alpha_{AB}dt, \tag{3.21}$$

in which $p_A^o(\tau)$ is the probability that the molecule remains in state A during the interval $[0, \tau]$.

To derive an expression for $p_A^o(t)$ consider that the fact that the molecule remains in state A means that it does not flip to state B. Hence, it obeys the following relation:

$$p_A^o(t + \Delta t) = p_A^o(t)(1 - \alpha_{AB}\Delta t).$$

After a little algebra this equation transforms into

$$\frac{p_A^o(t + \Delta t) - p_A^o(t)}{\Delta t} = -\alpha_{AB}p_A^O(t).$$

By taking the limit $\Delta t \to 0$ we finally get

$$\frac{dp_A^o(t)}{dt} = -\alpha_{AB}p_A^o(t), \tag{3.22}$$

whose solution is

$$p_A^o(\tau) = e^{-\alpha_{AB}\tau}. \tag{3.23}$$

By plugging Eq. (3.23) into Eq. (3.21) we finally obtain the probability distribution for the waiting times of one molecule flipping from A to B:

$$p_A(\tau)dt = \alpha_{AB} e^{-\alpha_{AB}\tau} dt, \tag{3.24}$$

which is nothing else but an exponential distribution with mean value $\overline{\tau}_{AB} = \alpha_{AB}^{-1}$. Therefore, the time the molecule remains in state A is a random variable obeying an exponential distribution. The average time the molecule remains in state A each time it gets there is α_{AB}^{-1}.

By following an analogous procedure we can show that the waiting-time probability distribution for state B is:

$$p_B(\tau)dt = \alpha_{BA} e^{-\alpha_{BA}\tau} dt. \tag{3.25}$$

The average waiting time in state B is $\overline{\tau}_{BA} = \alpha_{BA}^{-1}$.

Interestingly, from the average waiting times, we can compute the fraction of time that the system spends in states A and B as follows:

$$\pi_A = \frac{\tau_{AB}}{\tau_{AB} + \tau_{BA}} = \frac{\alpha_{BA}}{\alpha_{AB} + \alpha_{BA}},$$
$$\pi_B = \frac{\tau_{BA}}{\tau_{AB} + \tau_{BA}} = \frac{\alpha_{AB}}{\alpha_{AB} + \alpha_{BA}}.$$

These last results are in complete agreement with the previously calculated stationary probabilities of finding the molecule in states A and B—see Eqs. (3.13) and (3.14).

Knowing the probability distributions for the waiting times allows us to simulate the system stochastic evolution as follows:

1. Set the system initial state, A or B.
2. Set the initial time, $t = t_0$.
3. Compute the waiting time as a random number τ from an exponential distribution with mean value α_{XY}^{-1}, with X denoting the system current state.
4. Update the system state: if it is A change it to B, and vice versa.
5. Update the simulation time: $t = t + \tau$.
6. Iterate from step 3.

The above algorithm is an instance of the celebrated Gillespie algorithm. In the original paper (Gillespie 1977), the author proved that this algorithm renders exact stochastic simulations of the stochastic process described by its corresponding master equation—Eq. (3.10) in the present case. We implemented the previously described algorithm in `Python` and plotted the results of one of the simulations we carried out in Fig. 3.1. For such simulation we employed the following parameter values: $\alpha_{AB} = 1$ and $\alpha_{BA} = 2$.

Observe how the systems flips between the two states remaining a different time in each state every time it gets there. Nonetheless, despite this variability one can

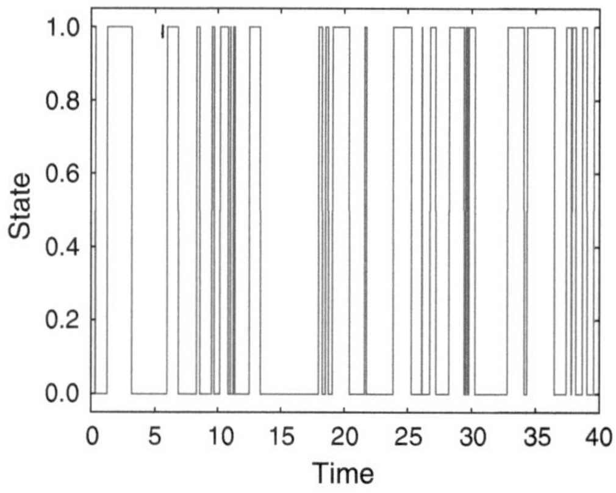

Fig. 3.1 Simulation of the random switching of a molecule between two states: A (0) and B (1)

appreciate that the molecule spends more time in state A than in state B. If the simulation is carried out for a very long time as compared with the average waiting times and one measures the total times the system spends in both states, one can verify that the molecule spends 2/3 of the total time in state A and one third in state B.

Consider now a system formed by N identical molecules flipping between states A and B. Let α_{AB} and α_{BA} be the propensities respectively associated with the switching of one molecule (if it were the only one in the system) from A to B and from B to A. The system state is determined by the number of molecules in A (n_A) and in B (n_B). From the above assertions, the propensity for the flipping of one of the n_A molecules in A to B is $\psi_{AB} = n_A\alpha_{AB}$. Conversely, the propensity for the flipping of one of the n_B molecules in B to A is $\psi_{BA} = n_B\alpha_{BA}$. Finally the propensity for the switching of one of the N molecules in the system to the opposite state is $\psi = \psi_{AB} + \psi_{BA}$.

From the considerations in the previous paragraph, and following a procedure analogous to the one leading to Eq. (3.24), one can demonstrate that the waiting time for the next switching (from whichever state) after the last one has taken place is a random variable obeying the following exponential distribution:

$$p(\tau)dt = \psi e^{-\psi\tau}dt. \tag{3.26}$$

Hence, the average waiting time between two consecutive switchings is

$$\bar{\tau} = \frac{1}{\psi} = \frac{1}{n_A\alpha_{AB} + n_B\alpha_{BA}}.$$

Interestingly, the waiting-time probability distribution, and hence the corresponding mean value, depend on the current state of the system.

Following Gillespie (1977) and taking into account the waiting time and the propensities computed above, the stochastic evolution of a system of N identical molecules flipping between states A and B can be simulated by means of the following algorithm:

1. Set the system initial state: $n_A = n_A^O$ and $n_B = n_B^O$.
2. Set the initial time, $t = t_0$.
3. Compute the propensities $\psi_{AB} = n_A \alpha_{AB}$, $\psi_{BA} = n_B \alpha_{BA}$, and $\psi = \psi_{AB} + \psi_{BA}$.
4. Randomly compute the waiting time τ from an exponential distribution with mean value ψ^{-1}.
5. Randomly choose whether one molecule flips from A to B or from B to A, considering that the first option has a probability equal to ψ_{AB}/ψ, while the probability of the second choice is ψ_{BA}/ψ.
6. Update the system state by modifying n_A and n_B according to the result of the previous step.
7. Update the simulation time: $t = t + \tau$.
8. Iterate from step 3.

We implemented the previously described algorithm in `Python` and carried out simulations with $N = 10, 100, 1{,}000$ molecules (the values of α_{AB} and α_{BA} are the same as those employed in the simulation of Fig. 3.1). The result is plotted in Fig. 3.2. Observe that, as expected the amount of noise decreases as the total number of molecules, N, increases. As a matter of fact, the stochastic simulations better approach the curves in Eqs. (3.5) and (3.6) as N gets larger. After a transient of less than 5 min, a stationary behavior is reached in which n_A and n_B fluctuate around the values given by Eqs. (3.7) and (3.8). Finally, as predicted by Eqs. (3.5) and (3.6), the time it takes the system to reach the stationary state is independent of the total molecule count.

3.4 Thermodynamics

Let us now analyze the chemical reaction from the perspective of thermodynamics. We have seen in Chap. 2 the so-called free energy change is the most informative thermodynamic quantity regarding chemical reactions. In particular, $\Delta G = 0$ when chemical equilibrium is concomitant with thermodynamic equilibrium.

According to Eq. (2.5), the free energy change for the reaction in (3.1) is

$$\Delta G = -\mu_A + \mu_B. \tag{3.27}$$

If we further take into consideration that

$$\mu_A = \mu_A^O + k_B T \ln N_A, \quad \mu_B = \mu_B^O + k_B T \ln N_B,$$

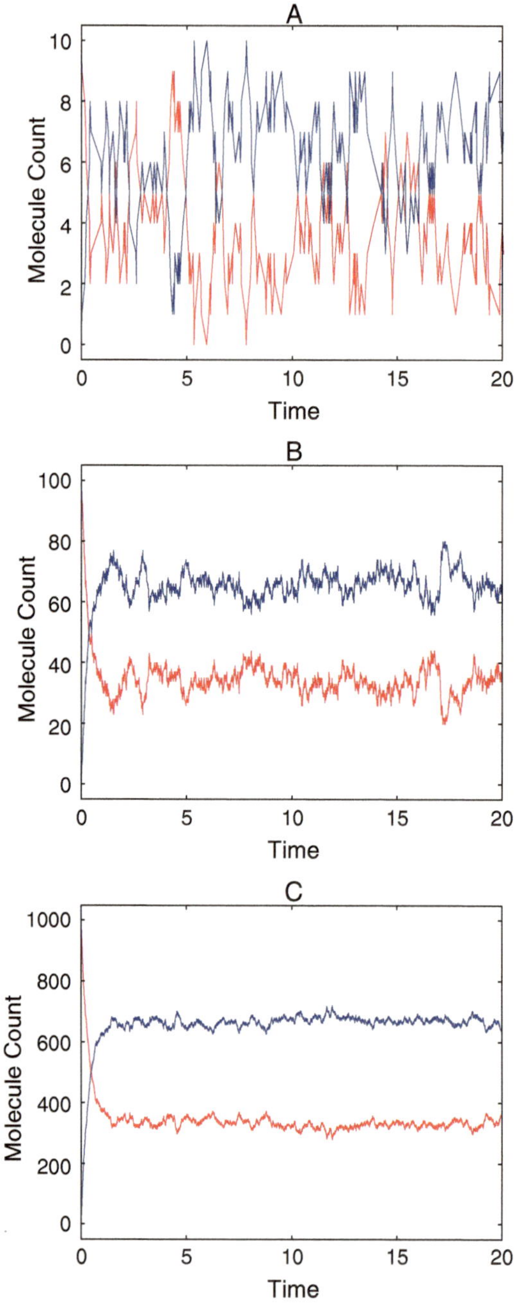

Fig. 3.2 Random simulation of the switching of N molecules between two states: A and B. The number of molecules in state A is plotted with *blue*, and the number of molecules in B is plotted with *red*. The plots in panels (**a**), (**b**), and (**c**) correspond to $N = 10, 100, 1{,}000$ respectively

It follows that the equilibrium assumption implies that

$$\frac{N_A^*}{N_B^*} = \frac{e^{\mu_B^O/k_B T}}{e^{\mu_A^O/k_B T}}, \tag{3.28}$$

where N_A^* and N_B^* denote the equilibrium molecule counts for species A and B, respectively. Observe that we have recovered the law of mass action. Moreover, a comparison with Eq. (3.9) suggests that $e^{\mu_A^O} \propto k_{AB}$ and $e^{\mu_B^O} \propto k_{BA}$. In the following section I shall informally prove the validity of this assertion, and the close relation between the three formerly analyzed approaches will become apparent along the proof.

3.5 The Fourth Musketeer, Linking Approaches

Consider once more a molecule that flips back and forth between states A and B. For example, think of a protein that shifts between two different conformational states. Strictly speaking this does not mean that A and B are the only states available for the protein. What it means is that, of the vast number of possible conformational states, only A and B are stable. The stability of a given state is determined by its energy, which in principle can be computed by adding up the interaction energies among all the molecule atoms. However, for the purpose of the present book we do not need a detailed picture of the resulting energy landscape, but only a rough sketch. In general, the energy landscape is a hyper-surface, which can be mathematically represented by a function of the form $\mathscr{E} = f(\mathbf{r})$, where $\mathbf{r}$ accounts for the position in the conformational space and $\mathscr{E}$ is the corresponding energy. In the present case, the fact that only two stable states exist, means that the energy landscape has two local minima—each local minima corresponding to one stable state (E and Vanden-Eijnden 2010)—divided by a separatrix. Then, if we lump together the conformational-state coordinates, the energy landscape for the reaction studied in the present chapter should look as sketched in Fig. 3.3.

In the absence of thermal agitation, the molecule state would evolve toward either one of the available stable states, and remain there indefinitely. However, at temperatures different from zero, the solvent molecules are constantly colliding with the molecule, perturbing it. Thus we can picture the molecule state as following a forced random walk in the conformational space: the interactions among the molecule atoms tend to make its state evolve toward a local minimum of the energy landscape, while the thermal perturbations tend to take it away from the local minima in a diffusive way.

Regard the energy landscape pictured in Fig. 3.3 and suppose that the protein is initially trapped in the basin of attraction surrounding state A. According to Kramers' theory (Van Kampen 1992; Risken 1996), the protein state will eventually escape from this basin due to thermal perturbations, and the escape rate

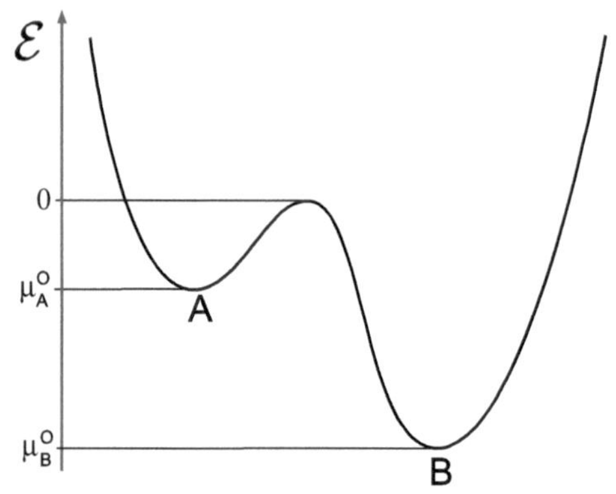

Fig. 3.3 Schematic representation of the energy landscape for a protein with two stable conformational states, *A* and *B*

(the probability per unit time that the protein state leaves the basin) is inversely proportional to the exponential of the energy-barrier height separating this basin from that of state B, divided by $k_B T$ (k_B being Boltzmann's constant). A similar argument applies in case that the molecule state is initially trapped in the basin surrounding state B. Let $\mathscr{E}_A$, $\mathscr{E}_B$, and $\mathscr{E}_C$ denote the energies of states A, B, and the separatrix between them (the saddle point separating the two basins of attractions). Assume without loss of generality that $\mathscr{E}_A < \mathscr{E}_B < \mathscr{E}_C = 0$. According to the discussion above, the escape rates from A to B, and vice versa, are:

$$\alpha_{AB} = \beta e^{\mathscr{E}_A/k_B T}, \tag{3.29}$$

$$\alpha_{BA} = \beta e^{\mathscr{E}_B/k_B T}, \tag{3.30}$$

with β a proportionality constant. In strict sense, the proportionality constant should be different for k_{AB} and k_{BA}. However no big error is introduced if we assume in a first approximation that they are equal.

With the aid of Eqs. (3.29) and (3.30), the relation between Eqs. (3.9) and (3.28) becomes clear. The chemical potential μ^O is nothing but the energy of a single molecule in solution located at the minimum of the basing of attraction surrounding the corresponding stable conformational state.

Let N_A and N_B denote the number of molecules whose states are in the basins surrounding states A and B, respectively. The probability per unit time (or propensity) that a molecule escapes from the basin surrounding A into that surrounding B is proportional to both N_A and $e^{\mu_A^O/k_B T}$. Since the chemical potential lumps these two variables into a single quantity, see Eq. (2.11), the propensity for the shift of a single molecule from A to B results to be proportional to $e^{\mu_A/k_B T}$. Similarly, the propensity for the shift of a single molecule from B to A happens to be proportional to $e^{\mu_B/k_B T}$. In other words, the chemical potential can be understood as minus the height of the *effective* energy barrier separating one given basin from the other. This effective barrier accounts for the number of molecules in each basin; the larger the molecule count, the smaller the effective barrier. The previous discussion further explains why the stationary state is reached when the chemical potential in both basins equilibrate (the effective energy barriers are the same for both states, and so the net fluxes from A to B and from B to A balance each other), and why the deeper the basin of a given state, the larger the number of molecules it has in the equilibrium.

As we have seen, the energy landscape approach is a common ground that links the three previously studied approaches: chemical kinetics, chemical master equation, and thermodynamics. Furthermore, the introduction of the energy landscape concept facilitates the realization that the three original approaches are nothing but different ways of modeling the same phenomenon. In that sense, every approach has particular advantages, but also disadvantages. What is then the most useful one? That depends on the kind of data we have and on the questions we are asking. In some cases, using more than one approach would be advisable. In any case, it is my claim that having a working knowledge of all the approaches is quite useful (and perhaps necessary). In that way we can choose the most suitable one according to the circumstances.

3.6 Sketching the Energy Landscape

Although conceptually very informative, in practice is quite difficult to have an accurate picture of the energy landscape. To do so, one would ought to compute the interaction energy of all the possible conformational states of the reacting molecules. Even if we had all the necessary information to perform such calculations (and we do not have it), the computational task would be impossible in most cases, even for the most powerful available computers. However, it is feasible to get a schematic representation of the energy landscape, which nonetheless has enough information for the present book purpose.

As stated above, in general, the energy landscape is a hyper-surface that can be mathematically represented by a function of the form $\mathscr{E} = f(\mathbf{r})$, where $\mathbf{r}$ accounts for the position in the conformational space and $\mathscr{E}$ is the corresponding energy.

For a given chemical reaction set, the energy landscape has as many local minima as the number of combinations of reactants and products participating in the reactions, and the height of each minima is determined by the sum of energies of the corresponding molecules. Consider for instance the reaction set

$$A + 2B \rightleftharpoons C,$$

$$A + C \rightleftharpoons D.$$

We see that chemical species D does not participate in the first reaction, while species B does not appear in the second one. However, they can be included without altering the nature of the chemical reactions as follows.

$$A + 2B + D \rightleftharpoons C + D,$$

$$A + B + C \rightleftharpoons B + D.$$

Then, we can assert that the energy surface has four local minima corresponding to $A + 2B + D$, $C + D$, $A + B + C$, and $B + D$, while the corresponding heights are $E_A + 2E_B + E_D$, $E_C + E_D$, $E_A + E_B + E_C$, and $E_B + E_D$, in which E_X denotes the energy of a molecule X in solution.

In the energy landscape, the minima are separated by separatrices whose height is given by the corresponding reaction-rate constants. The higher the separatrix between two minima, the smaller the reaction-rate constant of the connecting reaction. A non-existent reaction connecting two minima would then be represented by a separatrix of infinite height.

The principles summarized above are all that is needed to sketch the energy landscape of any reaction set, and were the ones employed to draw the energy landscape in Fig. 3.3. In the forthcoming chapters we shall invoke the same principles to schematically represent more energy landscapes in some particular cases. The process should become clearer as more examples are given.

3.7 Summary

In this chapter we introduced a third approach (besides chemical kinetics and thermodynamics) to study chemical reactions. This is a stochastic dynamics approach based on the chemical master equation. We also introduced the celebrated Gillespie algorithm to simulate the random evolution of a chemical reaction system. Rather than working with the most general reaction set, we used a very simple chemical reaction as an introductory example. With this example we were able to prove that the chemical kinetics approach is a particular case of the chemical master equation one. Moreover, by introducing the concept of energy landscape, we could see that the thermodynamics and the master equation approaches share a common ground. Thus, we could verify that the different approaches to studying a chemical reaction dynamics are nothing but different sides of a single dice. Each one has its own peculiarities and usefulness, but the underlying physics is the same for all of them. A detailed discussion about how to sketch the energy landscape for a particular reaction is given at the end of the chapter.

Chapter 4
Molecule Synthesis and Degradation

Abstract The present chapter is advocated to analyzing the dynamics of a simple birth–death process from the perspectives of chemical kinetics, stochastic processes, and thermodynamics. This process is important because, under certain conditions, it constitutes a good model for the numerous biomolecule production/degradation processes taking place within cells. To facilitate the achievement of the above stated goal, a couple of simple stochastic processes are previously introduced and analyzed in detail: the Poisson process and a simple pure death process. The objectives of the present chapter are twofold. On the one hand, we will gain deeper insight into the previous chapter methods, results, and conclusions, by tackling more elaborated examples. But also, the examples here introduced shall serve as building blocks for more realistic cellular-process models to be introduced thereafter. As a bonus, we also introduce and discuss some basic concepts of irreversible thermodynamics in the context of chemical kinetics.

4.1 Poisson Process

In every single living cell, the concentrations of all molecules remain more or less constant in time. As a matter of fact, this is a quintessential condition for homeostasis, which is a central concept in biology (Cannon 1929). These constant concentrations are achieved not because the whole system is static and no molecules are either produced or destroyed, but because the production and degradation rates are finely balanced. In the present chapter we shall study in detail the dynamics and thermodynamics of the simplest possible model accounting for molecule synthesis and degradation. But before going to the point, it is convenient to study the production and decay processes separately. Not only to get acquainted with them and thus facilitate the understanding of the combined process, but also to introduce some useful concepts.

M. Santillán, *Chemical Kinetics, Stochastic Processes, and Irreversible Thermodynamics*, Lecture Notes on Mathematical Modelling in the Life Sciences,
DOI 10.1007/978-3-319-06689-9_4, © Springer International Publishing Switzerland 2014

Consider a process consisting of a succession of independent events, each one having the same, constant, probability of occurrence per unit time, λ. In other words, the probability that one event happens in the interval $[t, t + \Delta t]$ is $\lambda \Delta t$, assuming that Δt is small enough so that none or at most one event takes place during such interval. Let $P(n, t)$ be the probability that n events have taken place up to time t. Then,

$$P(n, t + \Delta t) = P(n, t)(1 - \lambda \Delta t) + P(n - 1, t)\lambda \Delta t.$$

This equation accounts for the fact that there are only two possible ways in which n events occur up to time $t + \Delta t$: either the n events have already taken place up to time t and nothing happens in the interval $[t, t+\Delta t]$, or $n-1$ events have occurred up to time t and one more takes place during the interval $[t, t + \Delta t]$. After performing some algebra and taking the limit $\Delta t \to 0$, we obtain from the equation above that $P(n, t)$ obeys the following master equation

$$\frac{dP(n,t)}{dt} = \lambda \left(P(n-1, t) - P(n, t) \right). \tag{4.1}$$

The interested reader can verify without much trouble that the solution of Eq. (4.1) is

$$P(n, t) = \frac{(\lambda t)^n e^{-\lambda t}}{n!}, \tag{4.2}$$

which is nothing but a Poisson distribution with parameter λt (Evans et al. 2000). We have from the properties of the Poisson distribution the mean number of events occurring up to time t is

$$N(t) = \lambda t,$$

while the corresponding standard deviation is

$$\sigma_N(t) = \sqrt{\lambda t}.$$

From the above equations, the coefficient of variation for the event count is

$$CV_N(t) = \frac{\sigma_N(t)}{N(t)} = \frac{1}{\sqrt{\lambda t}}.$$

Observe that the coefficient of variation is a decreasing function of t. This means, that at very long times the Poisson process can be regarded as deterministic. We shall come back to this point later.

Another interesting property of the formerly defined stochastic process is the distribution of waiting times between consecutive events. Let $p(\tau)d\tau$ be the probability that the next event takes place in the infinitesimal interval $[\tau, \tau + d\tau]$, given that the previous one occurred at time $\tau = 0$. To compute this probability distribution take into consideration that $p(\tau)d\tau$ is the probability that no event takes place in the interval $[0, \tau]$, and that exactly one event occurs in the interval $[\tau, \tau + d\tau]$. That is,

$$p(\tau)d\tau = P(0, \tau)P(1, d\tau) = \lambda e^{-\lambda\tau}d\tau. \tag{4.3}$$

Hence, the waiting times obey an exponential distribution with mean λ^{-1} (Evans et al. 2000).

The stochastic process previously analyzed is known as the Poisson process (Ross 1983). As we have seen, it can be defined in three different, but equivalent, ways:

1. At most one event can occur in an infinitesimal time interval dt. This happens with probability λdt, independently of what happened in previous intervals.
2. The number of events n occurring in a finite interval of length t obeys the Poisson distribution given by Eq. (4.2).
3. The waiting times are independent and obey the exponential distribution in Eq. (4.3).

In agreement with the above definitions, the Poisson-process parameter, λ, deserves three different interpretations:

1. It is the probability per unit time that a single event occurs.
2. It measures the average number of events taking place per unit time; the rate of the process.
3. The average waiting time between consecutive events is λ^{-1}.

4.2 Biomolecule Synthesis as a Poisson Process

Under some conditions, a Poisson process is a good model for the synthesis of biomolecules. Think for instance of ATP or some other metabolite. If all the necessary substrates and enzymes are present at constant concentrations, then the metabolite synthesis rate is constant, and the production of individual molecules approaches a Poisson process. In a different example, transcription and translation are usually modeled as single-step processes (Shahrezaei and Swain 2008; Zeron and Santillán 2010). In particular, if transcription is not regulated—as for example in a constitutive promoter—then a Poisson process is a good model for the occurrence of successive transcriptional events.

Having a picture always helps to better understand a concept. With this purpose in mind, let us device an algorithm to simulate individual realizations of a Poisson process. The key for this algorithm is the waiting times between consecutive events; recall that the set of waiting times in a given realization of a Poisson process can be viewed as realizations of a random variable that obeys an exponential distribution with mean λ^{-1}. Once we have understood this, it is straightforward to device the

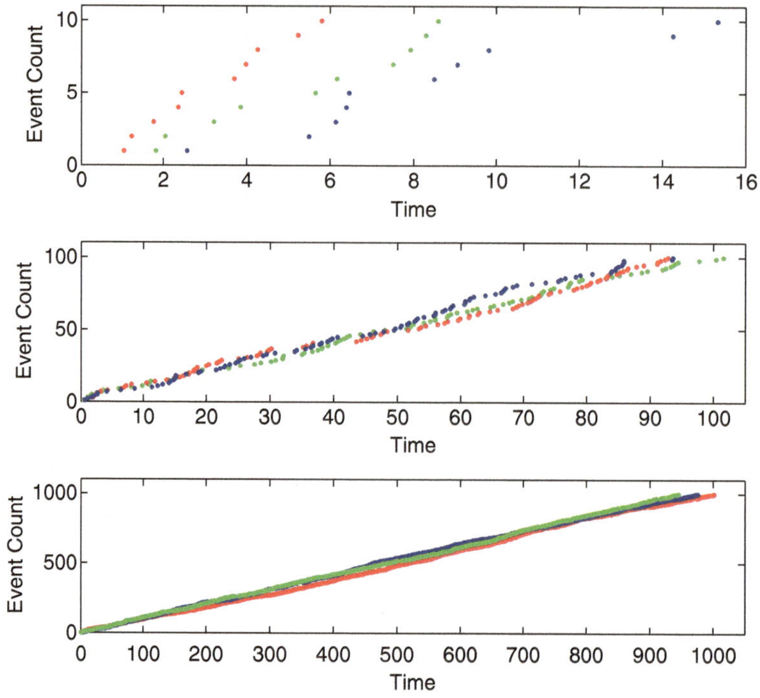

Fig. 4.1 Different realizations of a Poisson process, viewed at different time scales

following algorithm. Let n be an integer variable counting the number of events that have occurred, and t be a real variable accounting for the occurrence time of each individual event. The algorithm proceeds as follows:

1. Set $n = 0$ and $t = 0$.
2. Randomly compute the waiting time τ from an exponential distribution with mean λ^{-1}. One way of doing it is generating a uniform random number in the interval $[0, 1]$, r, and then calculating $\tau = -\log r/\lambda$.
3. Update $n = n + 1$ and $t = t + \tau$.
4. Iterate from step 2.

The results of computing different realizations of a Poisson process, using the algorithm above with $\lambda = 1$, are plotted in Fig. 4.1. Observe that when they are viewed at a time scale comparable to that of the mean waiting time, different realizations have quite different outcomes. However, when the observation time-scale is several orders of magnitude larger than λ^{-1}, then all the realizations are almost identical, and each of them approaches a continuous process taking place at constant rate. This is in agreement with the previous section assertion that a Poisson process can be regarded as deterministic at very large time scales, as compared with λ^{-1}. Finally, it is straightforward to prove that the average event count $N(t)$ obeys the following differential equation:

$$\frac{dN(t)}{dt} = \lambda,$$

which is what one gets for a deterministic process occurring at the constant rate λ.

Let us consider again a Poisson process as a model for transcription (gene expression). This process is known to take place at rates ranging in the order of a few to a few dozens per minute (McClure 1985). This implies that the average waiting time between consecutive transcriptional events is of the order of 1–10 s. Therefore, the deterministic approximation gives an accurate description when one counts the numbers of transcriptional events that take place in intervals of 20 min or more, for fast promoters, or in intervals of 3 h or more, for slow promoters. These intervals are relevant if we consider that the doubling times for the bacteria *E. coli* and the yeast *S. cerevisiae* are of the order of 30 min and 2 h, respectively. In other words, transcription can hardly be regarded in general as a deterministic process in the time life of unicellular organisms, and thus its stochastic nature cannot be ignored.

4.3 Exponential Decay

Consider a population of n molecules at time t and assume that the probability that each one of them decays in the interval $[t, t + \Delta t]$ is $\gamma \Delta t$. Under the assumption that all molecules decay independently from each other, the probability that one of the n molecules decays in the interval $[t, t + \Delta t]$ is $n\gamma\Delta t$, provided that Δt is small enough so that at most one molecule decays in such interval. Let $P(n, t)$ be the probability of having n molecules at time t. From the previous considerations, $P(n, t)$ obeys the following equation

$$P(n, t + \Delta t) = P(n + 1, t)(n + 1)\gamma\Delta t + P(n, t)(1 - n\gamma\Delta t).$$

To construct this equation we took into consideration that there are only to ways in which one can have n molecules at time $t + \Delta t$: either we had $n + 1$ molecules at time t and one decayed in the interval $[t, t+\Delta t]$, or we had n molecules at time t and none decayed during the interval $[t, t + \Delta t]$. After performing a little algebra and taking the limit $\Delta t \to 0$ we obtain the following master equation for the dynamics of $P(n, t)$ is

$$\frac{dP(n, t)}{dt} = \gamma(n + 1)P(n + 1, t) - \gamma n P(n, t). \tag{4.4}$$

The interested readers are invited to demonstrate that the general solution to this equation is

$$P(n, t) = \frac{\xi(t)^n e^{-\xi(t)}}{n!}, \tag{4.5}$$

with

$$\xi(t) = N_0 e^{-\gamma t}, \tag{4.6}$$

and N_0 denoting the initial average molecular count. We can see from Eq. (4.6) that the probability distribution describing this phenomenon is, once more, a Poisson distribution. However, in this case, the parameter of the distribution evolves as dictated by Eq. (4.6). We know from the properties of the Poisson distribution (Evans et al. 2000) that the mean and the variance are precisely $\xi(t)$. Let $N(t)$ and $\sigma_N(t)$ denote the average molecule count and the corresponding standard deviation. Then, from Eq. (4.6):

$$N(t) = \sigma_N^2(t) = N_0 e^{-\gamma t}. \tag{4.7}$$

The stochastic process just analyzed is a special case of the so-called death processes (Ross 1983). We call it exponential decay due to the fact that the average molecule count decreases exponentially and asymptotically goes to zero as $t \to \infty$.

Several interesting conclusions can be obtained from Eq. (4.7). First, the decay rate for $N(t)$ equals the probability of decay per unit time of individual molecules. Furthermore, the coefficient of variation evolves as

$$CV = \frac{\sigma_N(t)}{N(t)} = \frac{1}{\sqrt{N_0 e^{-\gamma t}}}.$$

Notice that CV is a function of both N_0 and time. Indeed, $CV = 1/\sqrt{N_0}$ at $t = 0$, and it increases monotonically and without limit as time passes. Assume that $N_0 \gg 1$. This means that $CV \ll 1$ for small times, $t \lesssim \gamma^{-1}$, and so the deterministic approximation $N(t) = N_0 e^{-\gamma t}$ provides an accurate description in the early stages of the decaying process. However, for times much larger than γ^{-1} the deterministic description is no longer good enough, and in consequence we are obliged to take into account the process stochastic nature.

Another way of corroborating the assertions in the previous paragraph is to look at the waiting times. Suppose that a decaying event occurred at time t and our system is left with n molecules. We are interested in the probability $p(\tau)d\tau$ that the next decaying event takes place in the interval $[t + \tau, t + \tau + d\tau]$. That is, no molecule must decay in the interval $[t, t + \tau]$, and one molecule should decay in $[t + \tau, t + \tau + d\tau]$. This means that $p(\tau)$ obeys the following equation (recall that the probability that one of the n molecules decays in an infinitesimal interval of length $d\tau$ is $\gamma n d\tau$):

$$p(\tau) = \left(1 - \int_0^\tau p(\tau')d\tau'\right)\gamma n.$$

By differentiating we obtain the following differential equation for $p(\tau)$:

$$\frac{dp(\tau)}{d\tau} = -\gamma n p(\tau).$$

The general solution of this differential equation is

$$p(\tau) = Ae^{-\gamma n \tau},$$

with A is an unknown constant whose value can be determined by imposing the normalization condition: $\int_0^\infty p(\tau)d\tau = 1$. After performing all the necessary calculations we get:

$$p(\tau) = \gamma n e^{-\gamma n \tau}. \tag{4.8}$$

We see from Eq. (4.8) that the waiting times obey an exponential probability distribution with parameter γn. Interestingly, this parameter not only depends on the decay-probability per unit time of individual molecules (γ), but also on the state of the system, which in this case is given by the number of existing molecules (n). We know from the properties of the exponential distribution (Evans et al. 2000) that the average waiting time and the corresponding standard deviation are both equal and given by $1/\gamma n$. This means that both the length and the variability of the waiting times increase as the number of molecules decrease, confirming that the description in terms of the mean $N(t)$ is only accurate when the system is composed of a considerably large number of molecules. Finally, it is straightforward to prove from Eq. (4.7) that $N(t)$ obeys the following differential equation

$$\frac{dN(t)}{dt} = -\gamma N(t), \tag{4.9}$$

which corresponds to the well-known deterministic differential equation employed to model exponential decay.

Knowing the distribution of waiting times allows us to simulate individual realizations of the decaying process by means of the Gillespie algorithm (Gillespie 1977). Let n be an integer variable representing the number of molecules in the system, and t be a real variable accounting for the occurrence time of individual degradation events. The algorithm goes as follows:

1. Set $n = n_0$ and $t = 0$.
2. Randomly compute the waiting time τ from an exponential distribution with mean $(\gamma n)^{-1}$. One way of doing it is generating a uniform random number in the interval [0, 1], r, and then computing $\tau = -\log r/\gamma n$.
3. Update $n = n - 1$ and $t = t + \tau$.
4. Iterate from step 2.

Three different simulations of decaying processes, carried out with the algorithm just described and $\gamma = 1$, are shown in Fig. 4.2. Notice that, as expected, all three of them are very much alike when the molecular count is large. As a matter of fact, the system evolution at large n values is well approximated by the deterministic description in terms of the average molecule count. However, this approximation is not good enough for low molecule counts. One fact worth of consideration is

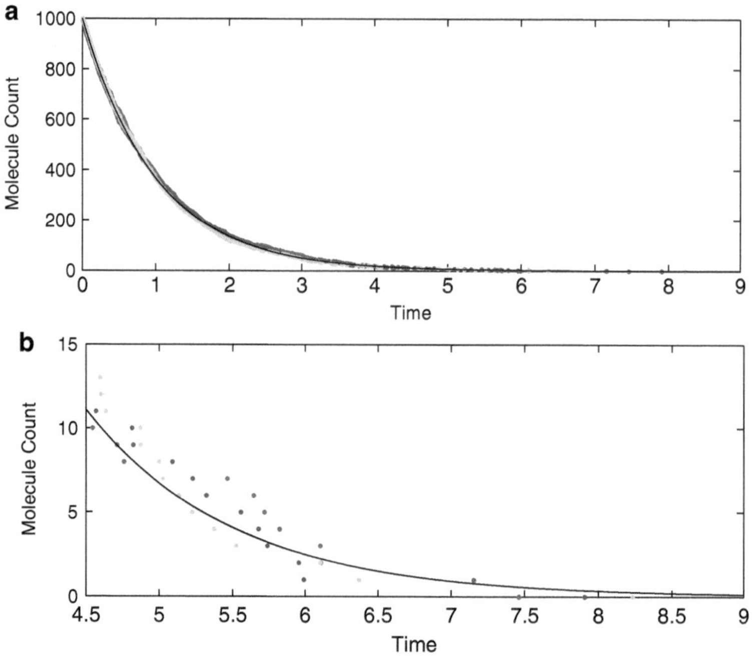

Fig. 4.2 Different realizations of an exponential decay process, all of them computed with $\gamma = 1$, together with the time evolution of the average molecule count (*black solid line*). (**a**) the whole processes. (**b**) zoom of the processes' last part

that the deterministic description predicts that the population never extinguishes completely, as the exponential function is never equal to zero. On the contrary, the stochastic simulations do predict a complete population extinction at a finite time. Nonetheless, the exact time of extinction is unpredictable (every simulation predicts a different extinction time, with a huge variability among them).

4.4 A Simple Birth–Death Process

Let us combine the two previous processes to develop a model for the dynamics of some biological molecules. Think for instance of ATP (or some other metabolite). Assume that all the substrates and enzymes involved in its synthesis are present at constant concentrations. This guaranties that molecule production takes place according to a Poisson process. On the other hand, ATP molecules are constantly hydrolyzed to deliver their energy cargo. If the probability per unit time that a molecule decays is constant along time (for instance, if the concentration of the hydrolyzing enzymes and of the products of hydrolysis are kept constant), then the decay process would be an exponential one. This model, is a special case of the so-called birth–death processes (Ross 1983). Although it comprises the two previously studies processes (Poissonian production and exponential decay), they cannot be studied separately because they are interlinked. To analyze the whole process we need to know the master equation for the probability of finding n molecules in the system at time t.

Let λ denote the probability per unit time that a new molecule is synthesized, and γ be the probability per unit time that a specific molecule decays. In other words, the probability that a new molecule is synthesized in an interval of length Δt is $\lambda \Delta t$, while the probability that one of the existing molecules decays in the same interval is $\gamma n \Delta t$ (where n is the current molecule count). From these considerations $P(n,t)$ obeys the following relationship

$$\begin{aligned} P(n, t+\Delta t) = {} & P(n-1,t)\lambda\Delta t(1-(n-1)\gamma\Delta t) \\ & + P(n+1,t)(1-\lambda\Delta t)(n+1)\gamma\Delta t \\ & + P(n,t)(1-n\gamma\Delta t)(1-\lambda\Delta t). \end{aligned}$$

The equation above accounts for the fact that ways through which we can have n molecules at time $t+\Delta t$ are as follows

1. There were $n-1$ molecules at time t, one was synthesized in the interval Δt and none was degraded in the same interval.
2. There were $n+1$ molecules at time t, no molecule was synthesized and one was degraded in the interval Δt.
3. There were n molecules at time t and neither a new molecule was produced nor one of the existing ones was degraded during the Δt period.

By doing some algebra and taking the limit $\Delta t \to 0$ the equation above transforms into:

$$\frac{dP(n,t)}{dt} = \lambda P(n-1,t) - \lambda P(n,t) + \gamma(n+1)P(n+1,t) - \gamma n P(n,t). \quad (4.10)$$

Before presenting the solution to Eq. (4.10) let us find the differential equation governing the dynamics of the mean value $N(t)$. By definition

$$N(t) = \sum_{n=0}^{\infty} n P(n,t).$$

The readers are invited to prove that, by differentiating the equation above, substituting Eq. (4.10), and using the normalization condition of the probability distribution $P(n,t)$ one gets

$$\frac{dN(t)}{dt} = \lambda - \gamma N(t). \quad (4.11)$$

To solve this differential equation let us make the following change of variable: $x(t) = N(t) - \lambda/\gamma$. In terms of variable x the differential equation in (4.11) becomes

$$\frac{dx(t)}{dt} = -\gamma x(t).$$

The general solution to this last equation is

$$x(t) = x_0 e^{-\gamma t}.$$

In the equation above x_0 is the initial value of $x(t)$. Revert now the change of variable to obtain

$$N(t) = \frac{\lambda}{\gamma} + \left(N_0 - \frac{\lambda}{\gamma}\right) e^{-\gamma t}, \tag{4.12}$$

with N_0 the initial value of $N(t)$. Observe from Eq. (4.12) that, regardless of the initial condition, $N(t)$ always converges to λ/γ as $t \rightarrow \infty$, and it does so exponentially. Moreover, although both λ and γ determine the stationary-state value of N, only γ influences the convergence velocity. The larger the value of γ, the faster $N(t)$ converges to λ/γ.

In the Poisson and the exponential decay processes the solution of the corresponding master equation was a Poisson distribution (with an adequate parameter), so it would not be a surprise if the solution to Eq. (4.10) is a Poisson distribution as well. Indeed, the readers won't have major problems to demonstrate by substitution that such solution is

$$P(n,t) = \frac{N(t)^n e^{-N(t)}}{n!}, \tag{4.13}$$

with $N(t)$ as given by Eq. (4.12). In other words, the solution to the master equation in Eq. (4.10) is a Poisson distribution that evolves in time following the mean value given by Eq. (4.12). Equation (4.13) further predicts the existence of a stationary solution:

$$P^s(n) = \frac{(\lambda/\gamma)^n \, e^{-\lambda/\gamma}}{n!}.$$

Interestingly, the value of the ratio λ/γ determines completely the shape and all other properties of the stationary distribution, but only γ influences the speed with which the master equation solution converges to it. Once more, the larger the value of γ the faster the convergence.

4.5 Gillespie Algorithm

We need an algorithm to simulate realizations of the production-decay process described in the previous section. As before, let us start by computing the distribution of waiting times. Assume that the system has n molecules at time t and let $p(\tau)d\tau$ be the probability that the next event (either the synthesis of a new molecule of the degradation of one of the existing ones) takes place in the interval $[t+\tau, t+\tau+d\tau]$. To compute this probability distribution we need to take into consideration that not only $p(\tau)d\tau$ is the probability that one event occurs in the interval $[t+\tau, t+\tau+d\tau]$, but also that none of them happens in the interval $[t, t+\tau]$. That is

$$p(\tau) = \left(1 - \int_0^\tau p(\tau')d\tau'\right)(\lambda + \gamma n).$$

Recall that $\lambda d\tau$ and $\gamma n d\tau$ are respectively the probabilities that a new molecule is synthesized and the one of the existing one is degraded in an interval of length $d\tau$. After differentiation, the last equation transforms into

$$\frac{dp(\tau)}{dt} = -(\lambda + \gamma n)p(\tau).$$

The general solution of the above differential equation is

$$p(\tau) = Ae^{-(\lambda+\gamma n)\tau},$$

with A a constant to be determined from the normalization condition for $p(\tau)$. After doing the corresponding calculations we obtain

$$p(\tau) = (\lambda + \gamma n)e^{-(\lambda+\gamma n)\tau}. \tag{4.14}$$

Therefore, the waiting times always obey an exponential distribution, but the corresponding parameter (and so the distribution properties) depends on the system state. Intriguingly, the parameter of the exponential distribution results to be the addition of the propensities (probabilities of occurrence per unit time) of the two possible reactions: molecule synthesis and degradation. Hence, we can define a state-dependent total propensity, a_T, equal to the sum of the individual propensities, a_i, and this total propensity determines the waiting-time distribution, given the system current state.

We have now all the necessary ingredients to develop an algorithm to simulate individual realizations of the production-decay process. Let n denote the number of molecules in the system and t a variable that records the time at which every individual chemical event takes place. Let us introduce as well new parameters ν_i accounting for the stoichiometry of the reactions taking place in the system (in our case, $\nu_1 = 1$ for the molecule synthesis reaction, and $\nu_2 = -1$ for the degradation reaction). Finally, let $a_1 = \lambda$ and $a_2 = \gamma n$ be the corresponding propensities. With the previous definitions, we can follow Gillespie (1977) to device the following algorithm:

1. Set $n = n_0$ (the initial molecule count), and $t = 0$.
2. Compute all a_i's according to the system state, as well as $a_T = \sum_i a_i$.
3. Calculate the waiting time by generating a random number from an exponential distribution with mean a_T^{-1}. For instance, this can be done by generating a uniform random number in the interval $[0, 1]$, r, and then computing $\tau = -\log r/a_T$.
4. Randomly choose which one of the two possible reaction takes place by assuming that the probability that reaction i happens is a_i/a_T. Suppose that we obtain as a result that reaction j is the one that will occur after the waiting time.
5. Update the system state according to the formerly computed waiting time and the chosen reaction: $t = t + \tau, n = n + \nu_j$.
6. Iterate from step 2.

The results of different simulations using this algorithm are presented in Fig. 4.3. In each simulation different values of λ and γ were employed, but they were chosen in such a way that the average molecular count is the same in all cases. Observe that, as expected, the molecular count fluctuates around the same average value ($\lambda/\gamma =$ 100 molecules) in all three simulations. Furthermore, the fluctuation amplitude is about the same in all cases. This was also expected because the coefficient of variation is always the same $CV = 1/\sqrt{\lambda/\gamma} = 0.1$. What changes are the fluctuation excursion times. The larger the value of γ, the faster the fluctuations disappear. This is in agreement with our previous finding that γ is the parameter that determines how fast the deterministic system evolves to the steady state, and how fast the probability distribution of the stochastic description (the solution of the master equation) converges to the stationary distribution.

4.6 Thermodynamic Interlude

Let us open a parenthesis to study the following set of chemical reactions from the perspective of chemical kinetics:

$$X \underset{N_A k_{AX}}{\overset{N_X k_{XA}}{\rightleftharpoons}} A \underset{N_Y k_{YA}}{\overset{N_A k_{AY}}{\rightleftharpoons}} Y, \tag{4.15}$$

assuming a constant total number of molecules $N_X + N_A + N_Y = N_T$. Given this last assumption, the present system bears no relation to the formerly introduced production-decay stochastic process. However, it will allow us to derive some important results which will be useful later on.

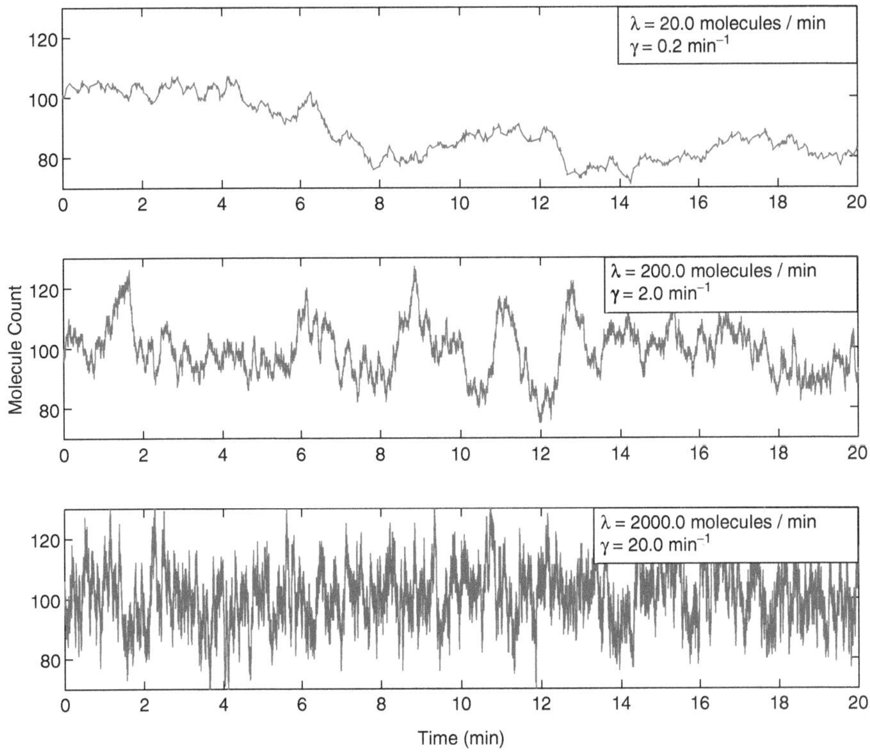

Fig. 4.3 Simulations of the production-decay process using the Gillespie algorithm and different values of the molecule synthesis and degradation rates

The set of differential equations governing the dynamics of the reactions in (4.15) is

$$\frac{dN_X}{dt} = -k_{XA}N_X + k_{AX}N_A,$$
$$\frac{dN_A}{dt} = k_{XA}N_X - k_{AX}N_A - k_{AY}N_A + k_{YA}(N_T - N_X - N_A).$$

A third equation, accounting for the dynamics of N_Y is not necessary because $N_Y = N_T - N_A - N_X$. For the time being we are not interested in the solution of this system of differential dynamics, but only in the steady-state behavior. In that regard, it is not hard to prove that the system has only one stable fixed point, and that it is given by

$$\overline{N}_A = \frac{1}{Q}N_T, \quad \overline{N}_X = \frac{1}{Q}\frac{k_{AX}}{k_{XA}}N_T, \quad \overline{N}_Y = \frac{1}{Q}\frac{k_{AY}}{k_{YA}}N_T,$$

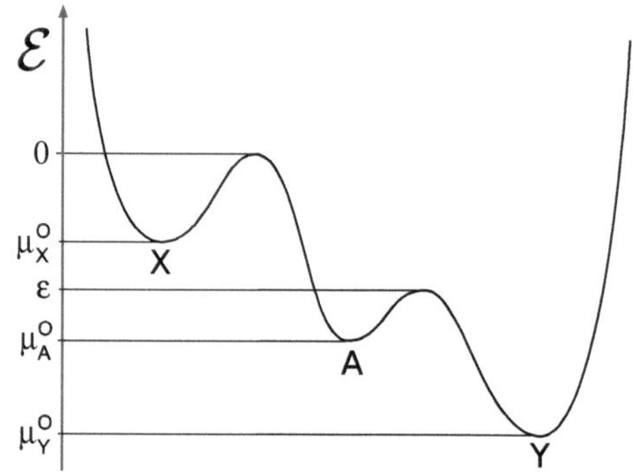

Fig. 4.4 Schematic representation of the energy landscape for a three-step chemical reaction

with

$$Q = 1 + \frac{k_{AX}}{k_{XA}} + \frac{k_{AY}}{k_{YA}}.$$

In particular, the above equations imply that

$$\overline{N}_X k_{XA} = \overline{N}_A k_{AX}, \quad \overline{N}_X k_{XB} = \overline{N}_B k_{BX}. \tag{4.16}$$

Let us sketch now the energy landscape for the chemical reactions in Eq. (4.15), following the principles delineated in Chap. 3. The result is presented in Fig. 4.4. From the results in Chap. 3 we know that

$$\begin{aligned} k_{XA} &= \beta e^{\mu_X^O/k_BT}, \\ k_{AX} &= \beta e^{\mu_A^O/k_BT}, \\ k_{AY} &= \beta e^{(\mu_A^O-\varepsilon)/k_BT}, \\ k_{YA} &= \beta e^{(\mu_Y^O-\varepsilon)/k_BT}. \end{aligned} \tag{4.17}$$

Substitution of these results into Eq. (4.16) leads to

$$\overline{N}_X e^{\mu_X^O/k_BT} = \overline{N}_A e^{\mu_A^O/k_BT} = \overline{N}_Y e^{\mu_Y^O/k_BT}.$$

Finally, by taking the logarithm and multiplying by k_BT we obtain

$$\mu_X^O + k_BT \ln \overline{N}_X = \mu_A^O + k_BT \ln \overline{N}_A = \mu_Y^O + k_BT \ln \overline{N}_Y = \mu_{eq}. \tag{4.18}$$

In other words, under the present conditions, the system steady state is reached when the chemical potential in all three states is the same—see Eq. (2.11). In the context of thermodynamics this situation corresponds to the so-called thermodynamic equilibrium (Lehninger et al. 2005).

Let me introduce a couple of additional closing remarks for the present section. The fact that all the chemical potentials reach the same value in the equilibrium, μ_{eq}, further implies that—see Eq. (4.18):

$$\overline{N}_X = \mathscr{A} e^{-\mu_X^O/k_B T}, \quad \overline{N}_A = \mathscr{A} e^{-\mu_A^O/k_B T}, \quad \overline{N}_Y = \mathscr{A} e^{-\mu_Y^O/k_B T},$$

with $\mathscr{A} = e^{\mu_{eq}/k_B T}$. This result is concomitant with the celebrated Boltzmann distribution for the system here studied. The second remark requires more calculations to be derived. From the results in Chap. 3, we can choose the Gibbs-free-energy zero level so that the free energy of a system held at constant pressure and temperature (like the present one) is

$$G = \sum_i N_i \mu_i.$$

In our case $i = X, A, Y$. After differentiating the above equation we get

$$dG = \sum_i (\mu_i + k_B T)\, dN_i,$$

recall that $\mu_i = \mu_i^O + k_B T \ln N_i$. If we take into account that in the steady state $\mu_i = \mu_{eq}$ for all i, and that $\sum_i N_i = N_T$, with N_T constant, it follows from the last equation that

$$dG|_{eq} = (\mu_{eq} + k_B T) \sum_i dN_i = 0. \tag{4.19}$$

In other words, the system Gibbs free energy is optimized in the equilibrium state. I leave for the reader to prove that, as a matter of fact, the Gibbs free energy is minimized in the equilibrium. This last result is a consequence of the second law of thermodynamics, which states that the Gibbs free energy minimizes in the equilibrium for systems held at constant pressure and temperature (Planck 1945).

4.7 Thermodynamic Interpretation of the Production-Decay Process

Consider once more the chemical-reaction system in (4.15) but assume that N_X and N_Y are constant. With this assumption, this system is equivalent to the previously studied production-decay process: molecules A are produced at a rate $N_X k_{XA} + N_Y k_{YA}$ and they are degraded at a rate $N_A(k_{AX} + k_{AY})$. Thus, the chemical-kinetics differential equation governing the dynamics of N_A is

$$\frac{dN_A(t)}{dt} = \lambda - \gamma N_A(t),$$

with $\lambda = N_X k_{XA} + N_Y k_{YA}$ and $\gamma = k_{AX} + k_{AY}$. Not surprisingly this last equation is identical to that in Eq. (4.11), and so its solution is given by (4.12). In particular, there exists one globally stable steady state:

$$\overline{N}_A = \frac{\lambda}{\gamma} = \frac{N_X k_{XA} + N_Y k_{YA}}{k_{AX} + k_{AY}}. \tag{4.20}$$

From the results in Chap. 1, the rates of the reactions in (4.15) are as follows:

$$J_{XA} = N_X k_{XA}, \quad J_{AX} = N_A k_{AX}, \quad J_{AY} = N_A k_{AY}, \quad J_{YA} = N_Y k_{YA}. \tag{4.21}$$

We see from this and Eq. (4.16) that the net fluxes between states X and A, and between states A and Y annihilate in the steady state of the system studied in the previous section. In contrast, if we substitute Eq. (4.20) into (4.21) we note that

$$J_{XA} \neq J_{AX} \quad \text{and} \quad J_{AY} \neq J_{YA}$$

in general, but

$$J = J_{XA} - J_{AX} = J_{AY} - J_{YA} = \frac{k_{XA} k_{AY} N_X - k_{AX} k_{YA} N_Y}{k_{AX} + k_{AY}}. \tag{4.22}$$

That is, there exists a nonzero net particle flux in the steady state of the present section system. However, the molecule count of state A remains constant because the influx equals the outflux. What happens then with the chemical potentials? To answer this question suppose again that the energy landscape determining the reaction rates is as in Fig. 4.4, and so that the reaction rates are given by Eq. (4.17). Thus, after substitution into Eq. (4.20) we get:

$$e^{\mu_A/k_B T} = \frac{e^{\mu_X/k_B T} + e^{-\varepsilon/k_B T} e^{\mu_Y/k_B T}}{1 + e^{-\varepsilon/k_B T}}. \tag{4.23}$$

Hence, in the present case the chemical potentials do not equate in the steady state, as they did in the previous-section example. This is in agreement with the existence of a net molecule flow because no flux could exist without a chemical-potential unbalance. Moreover, substitution of Eq. (4.17) into Eq. (4.22) allows us to compute the net molecule flux in terms of chemical potentials:

$$J = \beta \frac{e^{-\varepsilon/k_B T}}{1 + e^{-\varepsilon/k_B T}} \left(e^{\mu_X/k_B T} - e^{\mu_Y/k_B T}\right). \tag{4.24}$$

We see that the net molecule flux strongly depends on the chemical potentials of states X and Y, and is modulated by the difference of the energy landscape local maxima, $-\varepsilon$. Indeed,

$$\lim_{\epsilon \to -\infty} \frac{e^{-\varepsilon/k_B T}}{1 + e^{-\varepsilon/k_B T}} = 1,$$

while

$$\lim_{\epsilon \to \infty} \frac{e^{-\varepsilon/k_B T}}{1 + e^{-\varepsilon/k_B T}} = 0.$$

Finally, as the readers can easily verify, the Gibbs free energy does reach a constant value, but it does not minimize in the steady state. The reason being that the steady state does not correspond to chemical equilibrium in the present system. In agreement with this last assertion, one can define a stationary free energy flux given by

$$\phi = J(\mu_X - \mu_Y). \tag{4.25}$$

This flux accounts, on the one hand, for the net free energy flux entering the system due to molecules constantly being at state X and removed from state Y, to maintain the corresponding counts constant. On the other hand, ϕ also stands for the heat dissipated by the continuous molecule transition from a high energy state X to a low energy state Y (with an intermediate stop in A). In the long term, the system total free energy remains constant because the unbalance caused by the constant addition of high energy molecules and the removal at the same rate of low energy ones is compensated in the form of heat dissipation.

To conclude, we can assert that despite having a globally stable steady state, the birth–death process is a non-equilibrium phenomenon (thermodynamically speaking) because, in the steady state:

- There is a constant flow of molecules through the system.
- The stationary chemical potentials of the source, the intermediate, and the sink states are not equal.
- There is a constant input of free energy into the system associated with the molecule flow, and this energy input is balanced by heat dissipation.

Of course, these are not independent phenomena, but just different manifestations of the second law of thermodynamics (Beard and Qian 2008).

4.8 The General Form of the Chemical Master Equation and Gillespie Algorithm

In this and the last chapter we have derived the chemical Master equation, as well as Gillespie algorithm, for a few particular examples. Along the book we are going to employ both of them even more. Therefore, rather than deriving the particular forms each time, it is convenient to have a general derivation.

Consider a system that contains N different chemical species and in which M different chemical reactions are taking place. The state of such system can be denoted by a N-dimensional vector $\mathbf{x}$ such that its nth entry is a natural number equal to the number of molecules of the corresponding chemical species. Furthermore, each of the chemical reactions can be represented as

$$\mathbf{x} \xrightarrow{a_k(\mathbf{x})} \mathbf{x} + \boldsymbol{\nu}_k, \tag{4.26}$$

where $a_k(\mathbf{x})$ corresponds to the propensity of the kth reaction, while $\boldsymbol{\nu}_k$ is a N-dimensional vector containing the stoichiometric coefficients of the kth chemical reaction. That is, the nth entry of vector $\boldsymbol{\nu}_k$ determines how many molecules are created (or destroyed if it is a negative number) when an individual event of the kth chemical reaction takes place.

Let $P(\mathbf{x}, t)$ be the probability that the system is in state $\mathbf{x}$ at time t. Then, if we assume that Δt is small enough so at most one chemical event occurs, the probability that the system is in state $\mathbf{x}$ at time $t + \Delta t$ is given by:

$$\begin{aligned} P(\mathbf{x}, t + \Delta t) = &\sum_{k=1}^{M} P(\mathbf{x} - \boldsymbol{\nu}_k, t) a_k(\mathbf{x} - \boldsymbol{\nu}_k)\Delta t \prod_{k' \neq k} [1 - a_{k'}(\mathbf{x} - \boldsymbol{\nu}_k)\Delta t] \\ &+ P(\mathbf{x}, t) \prod_{k} [1 - a_k(\mathbf{x})\Delta t]. \end{aligned} \tag{4.27}$$

The first term on the right-hand side of Eq. (4.27) accounts for all the ways in which the system can evolve from a different state into state $\mathbf{x}$ through a single chemical event, while the second term takes into consideration the probability that the system remains in state $\mathbf{x}$ because no reaction takes place. By expanding the products and neglecting all the terms involving powers of Δt larger or equal than 2, Eq. (4.27) can be rewritten as

$$\begin{aligned} P(\mathbf{x}, t + \Delta t) \approx P(\mathbf{x}, t) + &\sum_{k=1}^{M} P(\mathbf{x} - \boldsymbol{\nu}_k, t) a_k(\mathbf{x} - \boldsymbol{\nu}_k)\Delta t \\ &- \sum_{k=1}^{M} P(\mathbf{x}, t) a_k(\mathbf{x})\Delta t. \end{aligned}$$

Finally, by rearranging terms and taking the limit $\Delta t \to 0$ we obtain

$$\begin{aligned}\frac{dP(\mathbf{x},t)}{dt} &= \lim_{\Delta t \to 0} \frac{P(\mathbf{x}, t+\Delta t) - P(\mathbf{x},t)}{\Delta t} \\ &= \sum_{k=1}^{M} P(\mathbf{x}-\mathbf{v}_k, t) a_k(\mathbf{x}-\mathbf{v}_k) - \sum_{k=1}^{M} P(\mathbf{x},t) a_k(\mathbf{x}). \end{aligned} \tag{4.28}$$

Equation (4.28) is the chemical master equation for the general chemical-reaction system in Eq. (4.26). As a matter of fact, Eq. (4.28) does not represent single equation but a system of equations; one equation for every available state. Given that in most cases the available-state count is very large (in no rare occasion it is indeed infinite), solving the chemical master equation system is extremely difficult, if not impossible. That is why having a way to numerically solve this equation is necessary. Perhaps the most popular algorithm to do so is the so-called Gillespie algorithm. Below we derive it.

Recall that a propensity is nothing but the probability per unit time than a given event occurs. Hence, if the propensity of the kth reaction when the system is in state $\mathbf{x}$ is $a_k(\mathbf{x})$, then the probability that one of the M different reactions takes place in an infinitesimal interval dt is

$$\sum_{k=1}^{M} a_k(\mathbf{x}) dt.$$

In other words, we can define a global propensity

$$a(\mathbf{x}) = \sum_{k=1}^{M} a_k(\mathbf{x}), \tag{4.29}$$

which can be interpreted as the probability per unit time that a single chemical step of one of the M possible reactions occurs. With this in mind, one can compute the probability distribution for the waiting times between two consecutive reactions as follows. Let $\mathbf{x}$ be the system state at time t, and let $p(\tau)d\tau$ denote the probability that the next reaction occurs in the interval $[t+\tau, t+\tau+d\tau]$. From its definition, $p(\tau)d\tau$ is the probability that no reaction happens in the interval $[t, t+\tau]$, and one of them occurs in the interval $[t+\tau, t+\tau+d\tau]$.. This assertion can be put into a mathematical equation as follows,

$$p(\mathbf{x},\tau) = \left(1 - \int_0^{\tau} p(\mathbf{x},\tau')d\tau'\right) a(\mathbf{x}).$$

By differentiating, this equation can be rewritten as

$$\frac{dp(\mathbf{x}, \tau)}{dt} = -a(\mathbf{x})p(\tau),$$

which is a differential equation whose general solution is

$$p(\mathbf{x}, \tau) = Ae^{-a(\mathbf{x})\tau},$$

where A is a constant to be determined from the normalization condition for $p(\tau)$. After performing the corresponding computations we obtain $A = a(\mathbf{x})$. Therefore, the probability distribution for the waiting time between consecutive chemical events comes out to be

$$p(\tau) = a(\mathbf{x})e^{-a(\mathbf{x})\tau}, \tag{4.30}$$

which is an exponential distribution whose parameter depends on the system state.

Once we know the probability distribution for the waiting times we can follow Gillespie (1977) to device the following algorithm to simulate the random evolution of chemical-reaction system in (4.26):

1. Set the initial time, $t = 0$, and the system initial state $\mathbf{x} = \mathbf{x}_0$.
2. Compute all the reaction propensities $a_k(\mathbf{x})$ given the current state, $\mathbf{x}$, and compute as well the global propensity $a(\mathbf{x}) = \sum_k a_k(\mathbf{x})$.
3. Calculate the waiting time until the next reaction by generating a random number from an exponential distribution with mean $a(\mathbf{x})^{-1}$. For instance, this can be done by generating a uniform random number in the interval $[0, 1]$, r, and then computing $\tau = -\log r/a(\mathbf{x})$.
4. Randomly choose which one of the possible reaction takes place by assuming that the probability that reaction k happens is $a_k(\mathbf{x})/a(\mathbf{x})$. Assume that the chosen reaction is denoted by index j.
5. Update the system state according to the formerly computed waiting time and chosen reaction: $t = t + \tau$, $\mathbf{x} = \mathbf{x} + \boldsymbol{\nu}_j$.
6. Iterate from step 2.

The above derived general forms of the chemical master equation and of Gillespie algorithm will be invoked in the following chapters to study some specific examples.

4.9 Summary

The main goal of the present chapter was to study the synthesis and degradation of molecules by means of the formerly reviewed approaches. To do so it was necessary to introduce and analyze (via the chemical master equation, as well as Gillespie algorithm) three novel stochastic processes: the Poisson process, the exponential decay process, and a very simple birth–death process. With the aid of these stochastic processes we were able to study the dynamic and thermodynamic aspects of molecule synthesis and degradation. Notably, we showed that when the synthesis and degradation rates equilibrate (with none of them being zero), a stationary state is reached, but it is not concomitant with thermodynamic equilibrium. This means that entropy is constantly being produced and heat is constantly being dissipated, despite the molecule count is constant in time. A side effect of our efforts to understand the above described phenomena was a deeper understanding of the chemical potential concept, which played a central role in this chapter. Finally, in the chapter's last section we derived the general forms of the chemical master equation, as well as Gillespie algorithm, for further reference.

Chapter 5
Enzymatic Reactions

Abstract In this chapter we generalize the birth–death process analyzed in the previous chapter to account for enzymatic molecule synthesis, rather than simple Poissonian production. To facilitate the analysis we assume a time-scale separation in the enzymatic reactions, and use it to reduce the complexity of the complete system. With this simplification the generalized birth–death process can be separated into two different subsystems that can be studied separately, and correspond to the systems studied in Chaps. 3 and 4. The simplification procedure, introduced in Sect. 5.1, is a very useful mathematical tool way beyond the scope of the present chapter.

5.1 Separation of Time Scales

Most biochemical reactions are catalyzed by enzymes. Therefore, although quite instructive, the model for the birth–death process studied in the previous chapter is not good enough an approximation in many instances. Typically, an enzymatic process consists of a series of chemical reactions that occur at different rates, and in some occasions it is possible to identify two well-separated time scales. When this occurs, the time-scale separation can be exploited to simplify the analysis of the whole system. Below we introduce a methodology to perform such simplification.

Consider a system in which molecules of N different chemical species are involved in M chemical reactions. The state of such a system is determined by the set of all the chemical-species molecule counts $\{n_1, n_2 \dots n_N\}$. Regardless of the value of N and the maximum values of $n_1, n_2 \dots n_N$, the set of all the possible system states is discrete, and in consequence the states can be enumerated. Thus, let us assume that $x = 1, 2 \dots$ labels of all the available states.

When an individual chemical event takes place, the molecule counts of some of the chemical species change, and so does the system state. Taking this into account,

M. Santillán, *Chemical Kinetics, Stochastic Processes, and Irreversible Thermodynamics*, Lecture Notes on Mathematical Modelling in the Life Sciences,
DOI 10.1007/978-3-319-06689-9_5, © Springer International Publishing Switzerland 2014

one can think of a general chemical master equation governing the system dynamics as follows:

$$\frac{dP(x;t)}{dt} = \sum_{x'} k(x';x)P(x';t) - k(x;x')P(x;t), \tag{5.1}$$

where $P(x;t)$ stands for the probability of finding the system in state x at time t, while $k(x;x')$ is the probability per unit time (also known as the propensity) that the system shifts from state x to state x'. Since the summation in Eq. (5.1) carried over all the possible values of x', we have implicitly assumed in its derivation that every pair of states x and x' are linked back and forth by chemical reactions. We know for a fact that this does not necessarily happen. But such supposition is not a problem because when the individual reaction that is supposed to link two given states does not exist, we simply take the corresponding propensity equal to zero.

Assume that a separation of time scales exists among the reactions taking place in the system. To account for it, let us suppose that the system states are labeled by means of a couple on integer variables (x, y) in such a way that the state transitions involving changes in y but not in x are orders of magnitude faster than all the others. The master equation for this system is a straightforward generalization of Eq. (5.1):

$$\frac{dP(x,y;t)}{dt} = \sum_{x',y'} k(x',y';x,y)P(x',y';t) - k(x,y;x',y')P(x,y;t). \tag{5.2}$$

In the equation above, the probabilities, P, and the propensities, k, have equivalent meanings as those in Eq. (5.1).

Let us rewrite Eq. (5.1) as

$$\begin{aligned}\frac{dP(x,y;t)}{dt} &= \sum_{x',y'} k(x,y';x,y)P(x,y';t) - k(x,y;x,y')P(x,y;t) \\ &+ \sum_{x' \neq x,y'} k(x',y';x,y)P(x',y';t) - k(x,y;x',y')P(x,y;t).\end{aligned}$$

Note that the first summation on the right-hand side of the above equation cancels because all the terms in it are added and subtracted once. Furthermore, if we sum over all y' values, taking into consideration that $P(x;t) = \sum_y P(x,y;t)$, and that $P(x,y;t) = P(x;t)P(y|x;t)$, we obtain

$$\frac{dP(x;t)}{dt} = \sum_{x'} \kappa(x';x,t)P(x';t) - \kappa(x;x',t)P(x;t), \tag{5.3}$$

with

$$\kappa(x';x,t) = \sum_{y,y'} k(x',y';x,y)P(y'|x';t).$$

Since the effective propensities $\kappa(x;x',t)$ are in terms of $P(y'|x';t)$, we need to know the dynamic behavior of these conditional probabilities. We have from the definition of conditional probability that

$$\frac{dP(y|x;t)}{dt} = \frac{1}{P(x;t)}\frac{dP(x,y;t)}{dt} - \frac{dP(x;t)}{dt}\frac{P(x,y;t)}{P(x;t)}.$$

After substituting Eqs. (5.2) and (5.3) into the last equation we obtain an expression in terms of the propensities $k(x',y';x,y)$ and $\kappa(x;x',t)$. However, due to the assumed separation of time scales, most of these propensities are negligible as compared with the ones accounting for changes in y but not in x (Santillán and Qian 2011). Hence, by only taking into account the terms containing $k(x,y';x,y)$ and considering that $p(y|x;t) = p(x,y;t)/p(x;t)$ we get

$$\frac{dP(y|x;t)}{dt} \approx \sum_{y'} k(x,y';x,y)P(y'|x;t) - k(x,y;x,y')P(y|x;t).$$

The time-scale separation can be invoked once more to make a quasi-stationary approximation consisting in the assumption that $P(y|x;t)$ instantaneously evolves to the stationary distribution, $P^s(y|x;t)$, given the current value of x. Therefore, from all the above considerations, the master equation that governs the dynamics of $P(x;t)$ results to be as follows:

$$\frac{dP(x;t)}{dt} = \sum_{x'} \kappa(x';x)P(x';t) - \kappa(x;x')P(x;t), \tag{5.4}$$

with

$$\kappa(x';x) = \sum_{y,y'} k(x',y';x,y)P^s(y'|x'), \tag{5.5}$$

while $P^s(y'|x')$ is the solution of

$$\sum_{y'} k(x,y';x,y)P^s(y'|x) - k(x,y;x,y')P^s(y|x) = 0. \tag{5.6}$$

In other words, we have a master equation that explicitly accounts for the dynamics of the slow variables, while the fast-variable dynamics is implicitly accounted for in the effective propensities $\kappa(x';x)$.

5.2 Enzymatic Production and Linear Degradation

Let us analyze the stochastic dynamics of a system in which A molecules are produced, via an enzymatic reaction, out of the substrate X, and are directly degraded into molecules Y. These processes can be summarized in the following set of chemical reactions (Lehninger et al. 2005; Houston 2001):

$$X + E \underset{k_{EX} \cdot n_{EX}}{\overset{k_{XE} \cdot n_X \cdot n_E}{\rightleftharpoons}} E_X, \tag{5.7}$$

$$E_X \underset{k_{AE} \cdot n_E \cdot n_A}{\overset{k_{EA} \cdot n_{EX}}{\rightleftharpoons}} E + A, \tag{5.8}$$

$$A \underset{k_{YA} \cdot n_Y}{\overset{k_{AY} \cdot n_A}{\rightleftharpoons}} Y. \tag{5.9}$$

In the above chemical reactions E stands for free enzyme; E_X is the enzyme–substrate complex; n_X, n_E, n_{EX}, n_A, and n_Y respectively represent the X, E, E_X, A, and Y molecule counts; and parameters k_{ij} denote reaction rates.

Under the assumptions that n_X and n_Y are constant, and that $n_E + n_{EX} = n_T$, with n_T constant, the system state is fully determined by the (n_{EX}, n_A) values. Let $P(n_{EX}, n_A; t)$ be the probability of having n_{EX} molecules E_X and n_A molecules A at time t. To study the system stochastic dynamics one could write the master equation for $P(n_{EX}, n_A; t)$ and analyze it. However, the analysis can be simplified if we previously make a quasi-stationary approximation. Is it usually acknowledged that the reaction in (5.7) is much faster than those in (5.8) and (5.9). Since the reaction in (5.7) modifies the value of n_{EX} but not of n_A, we can directly apply the formalism developed in the Sect. 5.1.

Let $P(n_A; t) = \sum_{n_{EX}=0}^{\infty} P(n_{EX}, n_A; t)$ be the probability of having n_A molecules A at time t. From Eq. (5.4) its dynamics are governed by

$$\begin{aligned}\frac{dP(n_A;t)}{dt} &= \kappa_{XA}(n_X, n_A - 1)P(n_A - 1; t) - \kappa_{XA}(n_X, n_A)P(n_A; t) \\ &\quad + \kappa_{AX}(n_X, n_A + 1)P(n_A + 1; t) - \kappa_{AX}(n_X, n_A)P(n_A; t) \\ &\quad + k_{AY}(n_A + 1)P(n_A + 1; t) - k_{AY} n_A P(n_A; t) \\ &\quad + k_{YA} n_Y P(n_A - 1; t) - k_{YA} n_Y P(n_A; t),\end{aligned} \tag{5.10}$$

in which, according to Eq. (5.5), $\kappa_{XA}(n_Y, n_A)$ and $\kappa_{AK}(n_Y, n_A)$ are given by

$$\kappa_{XA}(n_X, n_A) = \sum_{n_{EX}} k_{EA} n_{EX} P^s(n_{EX}|n_A) = k_{EA}\overline{N}_{EX}, \tag{5.11}$$

$$\kappa_{AX}(n_X, n_A) = \sum_{n_{EX}} k_{AE}(1 - n_{EX})n_A = k_{AE}\left(1 - \overline{N}_{EX}\right)n_A, \tag{5.12}$$

with

$$\overline{N}_{EX} = \sum_{n_{EX}} n_{EX} P^s(n_{EX}|n_A). \tag{5.13}$$

Regarding $P^s(n_{EX}|n_A)$, according to Eq. (5.6) it is the stationary solution of the following master equation:

$$\begin{aligned}\frac{dP(n_{EX}|n_A;t)}{dt} &= k_{XE}n_X(n_T - n_{EX} + 1)P(n_{EX} - 1|n_A;t)\\ &\quad - k_{XE}n_X(n_T - n_{EX})P(n_{EX}|n_A;t)\\ &\quad + k_{EX}(n_{EX} + 1)P(n_{EX} + 1|n_A;t)\\ &\quad - k_{EX}n_{EX}P(n_{EX}|n_A;t).\end{aligned} \tag{5.14}$$

A comparison of Eqs. (5.14) and (3.17) reveals that they are equivalent, and so that the process modeled by Eq. (5.14) is that of n_T molecules flipping between states E and E_X, with constant transition rates for individual molecules. Thus, from the results in Chap. 3—Eqs. (3.13), (3.14), (3.18), and (3.19)—the stationary distribution $P^s(n_{EX}|n_A)$ happens to be a binomial distribution with parameters $n = n_T$ and $p = k_{XE}n_X/(k_{XE}n_X + k_{EX})$. Moreover, the average number of molecules in the states E_X and E are given by

$$\overline{N}_{EX} = n_T \frac{k_{XE}n_X}{k_{XE}n_X + k_{EX}} = n_T \frac{n_X}{n_X + K_E}, \tag{5.15}$$

$$\overline{N}_E = n_T - \overline{N}_{EX} = n_T \frac{K_E}{n_X + K_E}, \tag{5.16}$$

with $K_E = k_{EX}/k_{XE}$. Finally, after substituting back into Eq. (5.10) we obtain the following master equation for $P(n_A, t)$:

$$\begin{aligned}\frac{dP(n_A;t)}{dt} &= k_{EA}\overline{N}_{EX}P(n_A - 1;t) - k_{EA}\overline{N}_{EX}P(n_A;t)\\ &\quad + k_{AE}\overline{N}_E(n_A + 1)P(n_A + 1;t) - k_{AE}\overline{N}_E n_A P(n_A;t)\\ &\quad + k_{AY}(n_A + 1)P(n_A + 1;t) - k_{AY}n_A P(n_A;t)\\ &\quad + k_{YA}n_Y P(n_A - 1;t) - k_{YA}n_Y P(n_A;t),\end{aligned} \tag{5.17}$$

where $\overline{N}_{EX}$ and $\overline{N}_E$ are given by Eqs. (5.15) and (5.16).

Notice that Eq. (5.17) is the same as Eq. (4.10). Hence, the quasi-stationary approximation, allowed us to reduce the system of chemical reactions in (5.1) to a birth–death process in which the effective production and degradation rates are:

$$\lambda = k_{EA}\overline{N}_{EX} + k_{YA}, \tag{5.18}$$

$$\gamma = k_{AE}\overline{N}_{E} + k_{AY}. \tag{5.19}$$

This further implies that $P(n_A, t)$ obeys a Poisson distribution whose parameter converges exponentially (with rate γ) to the stationary value λ/γ.

To investigate the dynamics of the average number of A molecules define $N_A(t) = \sum_{n_A=0}^{\infty} n_A P(A;t)$. By following the procedure leading to Eq. (4.11), we get the following differential equation:

$$\frac{dN_A(t)}{dt} = \left(k_{EA} n_T \frac{n_X}{n_X + K_E} + k_{YA}\right) - \left(k_{AE} \frac{K_E}{n_X + K_E} + k_{AY}\right) N_A(t). \tag{5.20}$$

The enzymatic and the degradation reactions are usually regarded as irreversible. Although this is not strictly possible because all chemical reactions are reversible (Lehninger et al. 2005), it might happen that the backward reactions occur with rates so small that they can be consider as negligible from the standpoint of chemical kinetics. If this is the case then $k_{AE}, k_{YA} \approx 0$, and

$$\frac{dN_A(t)}{dt} \approx k_{EA} n_T \frac{n_X}{n_X + K_E} - k_{AY} N_A(t). \tag{5.21}$$

Interestingly, Eq. (5.21) is the differential equation commonly employed in the context of chemical kinetics to model the dynamics of a chemical species that is produced via a catalytic reaction and degraded linearly (Houston 2001). In particular, the first term on the right-hand side of Eq. (5.21) corresponds to Michaelis–Menten equation, which is commonly used to model the velocity of enzymatic reactions (Houston 2001; Lehninger et al. 2005). Let us close this section by stating that knowing how Eq. (5.21) can be deduced from a stochastic chemical dynamics approach, allows us to better understand its range of validity and its connection with the relevant quantities of the stochastic-description.

5.3 Thermodynamics of Enzymatic Reactions

The quasi-stationary approximation studied in the previous section allowed us to break the system of chemical reactions in (5.7)–(5.9) into a couple of subsystems that can be analyzed separately:

1. The interaction of the enzyme E with the substrate X to form the complex E_X.
2. The synthesis of molecules A from E_X, and their degradation into molecules Y.

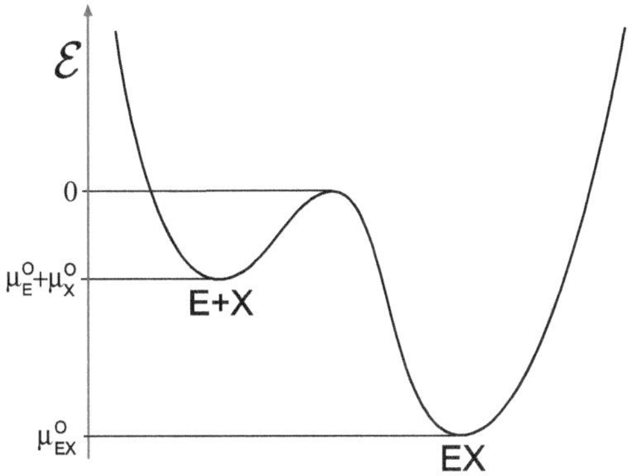

Fig. 5.1 Schematic representation of the energy landscape for the reaction $E + X \rightleftharpoons E_X$

If we view the whole system from a slow dynamics perspective (that of A molecules), we can think of subsystem 1 as having so fast dynamics that it equilibrates instantaneously with the slow changing variables. At the same time, a second subsystem (labeled 2), in which the effects of the fast subsystem are averaged over the fast time scale, becomes apparent. Subsystem 2 corresponds to the synthesis and degradation of A molecules.

The forthcoming discussion relies upon the assumption that the system is observed at a slow time scale. From this perspective, subsystem 1 corresponds to the chemical reaction analyzed in Chap. 3. That is, n_T enzyme molecules flip between states E and E_X. The rate with which individual molecules shift to state E_X is $k_{XE}n_X$. This rate is constant because we have assumed a constant n_X. On the other hand, the rate with which an individual enzyme in state E_X flips to state E is k_{EX}. Note that the reaction $E_X \rightleftharpoons E + A$ is not taken into account. The reason being that in the fast time scale it barely takes place. Finally, from the slow dynamics perspective, subsystem 1 can always be considered to be in a stationary state given the current n_X value.

From the discussion in the previous paragraph and the results in Chap. 3 we can assert that the probability distribution of having n_{EX} of the n_T enzymes in state E_X is a binomial distribution in which the success probability—the probability that a single enzyme is in state E_X—is $n_X/(n_X + K_E)$.

From the point of view of thermodynamics, the stationary distribution of enzyme molecules between states E and E_X is concomitant with thermodynamic equilibrium. To understand this, let us assume that the average energies of a free enzyme, a free molecule X, and the complex E_X are $\mathscr{E}_E$, $\mathscr{E}_X$, and $\mathscr{E}_{EX}$, respectively. Further assume that, in order for an enzyme and a molecule X to bind, their combined energies need to surpass an energy threshold $\mathscr{E}_{th} > \mathscr{E}_E, \mathscr{E}_X, \mathscr{E}_{EX}$, which can be assumed zero without loss of generality. Thus, the energy landscape for this reaction will look like the one sketched in Fig. 5.1, while the scape rates from $E + X$ to E_X and vice versa are

$$k_{XE} = \beta e^{(\mathscr{E}_E+\mathscr{E}_X)/k_BT}, \quad k_{EX} = \beta e^{\mathscr{E}_{EX}/k_BT}. \tag{5.22}$$

On the other hand, we can see from (5.15) and (5.16) that the condition for the stationary state is

$$k_{XE} n_X \overline{N}_E = k_{EX} \overline{N}_{EX}. \tag{5.23}$$

Henceforth, by substituting (5.22) into (5.23) we obtain

$$e^{\mathscr{E}_E/k_BT}\overline{N}_E e^{\mathscr{E}_X/k_BT} n_X = e^{\mathscr{E}_{EX}/k_BT}\overline{N}_{EX}.$$

Finally, by taking the logarithm and multiplying by k_BT on both sides, the above equation becomes

$$\mu_E + \mu_X = \mu_{EX}, \tag{5.24}$$

where $\mu_i = \mu_i^O + k_BT \ln n_i$ and $\mu_i^O = \mathscr{E}_i$ $(i = E, X, EX)$. In other words, the sum of the chemical potentials of the substances on both sides of the reaction $E + X \rightleftharpoons E_X$ is the same, implying thermodynamic equilibrium (Lehninger et al. 2005).

Regarding subsystem 2, it is equivalent to the birth–death process studied in Chap. 4. A molecules are synthesized out of the complex E_X and are degraded into Y molecules. We have seen that the E_X molecule count is not constant but fluctuating, and that the fluctuations obey a binomial distribution. However, from the standpoint of A-molecule dynamics, only the average $\overline{N}_{EX}$ value matters and the fluctuations can be disregarded. The reason for this is the assumed separation of time scales: what is "instantaneous" in the slow dynamics perspective (that of A molecules) involves a "long" time period in the time scale of the reaction $E + X \rightleftharpoons E_X$. This means that while one A molecule is produced or degraded, n_{EX} fluctuates several times. However, from the point of view of A molecules, it is impossible to follow individual fluctuations and only the average effect is felt.

Since the production and decay events of A molecules are stochastic, n_A fluctuates around the average value $N_A(t)$ and these fluctuations obey a Poisson distribution. On the other hand, $N_A(t)$ evolves with time according to Eq. (4.12). This further means that $N_A(t)$ exponentially converges to

$$\overline{N}_A = \frac{\lambda}{\gamma},$$

with λ and γ as given by Eqs. (5.18) and (5.19). We can finally assert from the results in Chap. 4 that, once the system has reached the stationary state, the average duration of individual fluctuations is determined by the value of γ solely: the larger the γ the shorter the fluctuations.

In the stationary state, the average production and decay rates of A molecules are constant and equal, so $\overline{N}_A$ remains unchanged. If we assume that the production

Fig. 5.2 Schematic representation of the energy landscape for the slow subsystem of the reaction scheme (5.7)–(5.9)

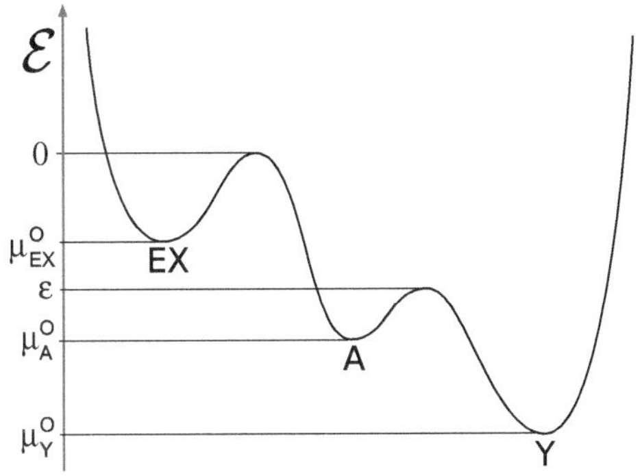

and degradation of A molecules are irreversible reactions, the value of both rates is given by—see Eq. (4.22):

$$J = k_{EA}\overline{N}_{EX} = k_{AY}\overline{N}_A = k_{EA}n_T\frac{n_X}{n_X + K_E}. \tag{5.25}$$

Following with the comparison with the birth–death process of Chap. 4, we can think of an energy landscape for subsystem 2 as depicted in Fig. 5.2. As in the birth–death process of Chap. 4, the system does evolve to a stationary state, but such stationary state does not correspond to chemical equilibrium. This happens because the stationary molecular counts do not eliminate the original chemical-potential unbalance, and so $\mu_{EX} \neq \mu_A \neq \mu_Y$ in general. Concomitantly, there exists a constant influx of energy (originated by the input of high energy molecules and the output of low energy molecules), which in turn is dissipated as heat. Under the supposition that the production and degradation of A molecules are irreversible reactions, the stationary energy influx and heat dissipation rates are—see Eq. (4.25):

$$\phi = J(\mu_{EX} - \mu_Y) = k_{EA}n_T\frac{n_X}{n_X + K_E}(\mu_{EX} - \mu_Y). \tag{5.26}$$

We see from Eq. (5.25) that we could either increase the amount of enzymes of augment the number of substrate molecules (n_X) to accelerate the enzymatic reaction. However, this also increases the heat dissipation rate (5.25). As a matter of fact, the only way to stop heat dissipation (and so making the enzymatic reaction thermodynamically reversible) is to make the reaction velocity equal to zero.

5.4 Summary

This chapter generalizes the results in Chap. 4. Here we study molecule synthesis and degradation one more time, but take into consideration the case in which the synthesis process is catalyzed by an enzyme. To simplify the chemical master equation that models the stochastic evolution of this system we introduced a very useful technique that makes use of a naturally existing separation of time scales. With this, we could split the system into a couple of nested subsystems which correspond to the processes studied in Chaps. 3 and 5. Thus the conclusions obtained there can be extrapolated to the system here studied. In particular, we could understand how enzymes accelerate chemical reactions without altering the system thermodynamic behavior.

Chapter 6
Receptor–Ligand Interaction

Abstract In this chapter we study the dynamic and thermodynamic behavior of a ubiquitous biochemical process: the interaction of a receptor molecule with multiple ligands. As before, we study it from the perspectives of chemical kinetics, stochastic processes, and thermodynamics. In essence we do not introduce any new tools, but make use of the ones introduced in the previous chapters. However, as the readers will be aware by the end of the chapter, the obtained conclusions allow a deeper understanding not only of this particular system but also of the dynamic and thermodynamic behavior of biochemical systems in general. It is our goal that, by gradually increasing the complexity of the studied systems, while building up upon previous examples, the reader will develop a more profound notion of the way the different approaches are inter-related. In particular, the present chapter includes a thorough discussion about the thermodynamic behavior of the studied system and how it compares with the ones in previous chapters.

6.1 Ligand–Receptor Interaction

So far we have studied three different sets of chemical reaction schemes. We started in Chap. 3 with a system in which the stationary state corresponds to thermodynamic equilibrium, followed by two systems (Chaps. 4 and 5) that never reach thermodynamic equilibrium, even in the steady state. Interestingly, in the system studied in Chap. 3, a constant number of molecules (no molecules are either produced or degraded) switches between two different states, while in the systems of Chaps. 4 and 5, an average molecule count is maintained because molecule production and degradation balance each other. We can think of the systems in Chaps. 4 and 5 as involving a constant flux of molecules that are synthesized out of a source and are degraded into a sink. This suggests that: as long as there is a null flux of molecules in a chemical reaction system, its stationary state is compatible with thermodynamic equilibrium. In the present chapter we introduce another system that complies with this generalization. In it, a receptor molecule R is bound in two different sites by

M. Santillán, *Chemical Kinetics, Stochastic Processes, and Irreversible Thermodynamics*, Lecture Notes on Mathematical Modelling in the Life Sciences, DOI 10.1007/978-3-319-06689-9_6,

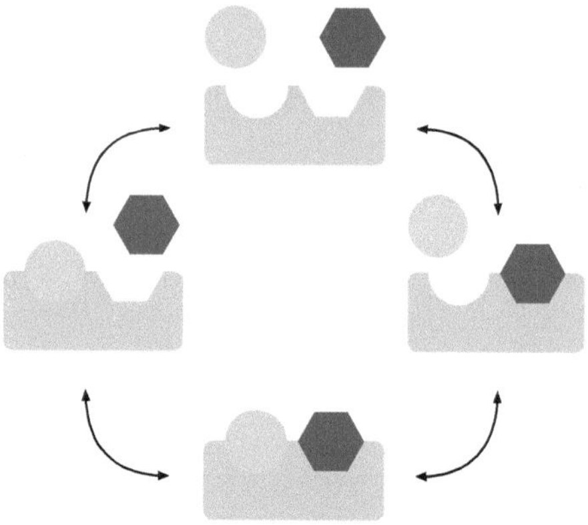

Fig. 6.1 Schematic representation of the reactions undergone by a receptor that can be bound by two different ligands in two different binding sites

ligand molecules X and Y, as schematically represented in Fig. 6.1. This system is representative of many processes taking place within living cells. To quote a few examples: a DNA promoter that needs to be bound by various transcription factors and an RNA polymerase in order to start transcription, an ion channel whose gating is regulated by the binding of two or more regulatory molecules, and an enzyme that is inhibited when it is bound by multiple ligands at different sites. Although the present chapter is limited to studying the case in which a receptor is bound by two ligands, it is straightforward to extend the analysis to the binding of an arbitrary number of ligands.

6.2 Stochastic Kinetic Analysis of a Single Receptor Molecule

Consider a molecule (called the receptor and denoted by R) that has two binding sites, one specific for X molecules and the other specific for Y molecules. The processes through which the binding sites are occupied and unoccupied are schematically represented in Fig. 6.1. Let us assume for the moment that the binding of a molecule X to its corresponding binding site is not influenced by the state of the Y binding site, and vice versa. Under this assumption, the processes illustrated in Fig. 6.1 can be conceptualized in terms of the following set of chemical reactions:

$$R + X \underset{k_X^- n_{RX}}{\overset{k_X^+ n_R n_X}{\rightleftharpoons}} R_X, \tag{6.1}$$

$$R + Y \underset{k_Y^- n_{RY}}{\overset{k_Y^+ n_R n_Y}{\rightleftharpoons}} R_Y, \tag{6.2}$$

$$R_X + Y \underset{k_Y^- n_{RXY}}{\overset{k_Y^+ n_{RX} n_Y}{\rightleftharpoons}} R_{XY}, \tag{6.3}$$

$$R_Y + X \underset{k_X^- n_{RXY}}{\overset{k_X^+ n_{RY} n_X}{\rightleftharpoons}} R_{XY}. \tag{6.4}$$

In the reactions above, k_X^+ and k_X^-, respectively, represent the propensities (probability per unit time) for the binding and unbinding of a molecule X to its corresponding site; k_Y^+ and k_Y^- are the corresponding propensities (also known as reaction rate constants) for the binding and unbinding of a molecule Y; n_X and n_Y are, respectively, the molecule counts of ligands X and Y; and n_R, n_{RX}, n_{RY}, and n_{RXY} correspond to the number of receptor molecules that are free, bound by a ligand X, bound by a ligand Y, and bound by both ligands, respectively.

Let us consider one single receptor molecule, which when bound and unbound by ligands X and Y flips among states $(0,0)$, $(X,0)$, $(0,Y)$, and (X,Y). If n_X and n_Y are assumed constant, the master equation (actually the master-equation family) describing the dynamics of the probability distribution for this system is

$$\frac{dP_{00}(t)}{dt} = k_X^- P_{X0}(t) - k_X^+ n_X P_{00}(t) + k_Y^- P_{0Y}(t) - k_Y^+ n_Y P_{00}(t), \tag{6.5}$$

$$\frac{dP_{X0}(t)}{dt} = k_X^+ n_X P_{00}(t) - k_X^- P_{X0}(t) + k_Y^- P_{XY}(t) - k_Y^+ n_Y P_{X0}(t), \tag{6.6}$$

$$\frac{dP_{0Y}(t)}{dt} = k_Y^+ n_X P_{00}(t) - k_Y^- P_{0Y}(t) + k_X^- P_{XY}(t) - k_X^+ n_Y P_{0Y}(t), \tag{6.7}$$

$$\frac{dP_{XY}(t)}{dt} = k_Y^+ n_Y P_{X0}(t) - k_Y^- / k_C P_{XY}(t) + k_X^+ n_X P_{0Y}(t) - k_X^- P_{XY}(t), \tag{6.8}$$

in which $P_{ij}(t)$ stands for the probability that the receptor molecule is in state (i, j) at time t.

Although it is possible to directly solve Eqs. (6.5)–(6.8), it is easier if we take advantage of the independence of the X and Y binding sites to split this system of coupled differential equations into two simpler independent master equations. Let $P_X(t)$ $(P_Y(t))$ denote the probability that the X (Y) binding site is bound by the corresponding ligand at time t, regardless of the state of the other binding site. Since both sites are independent, the master equations for $P_X(t)$ and $P_Y(t)$ should be

$$\frac{dP_X(t)}{dt} = k_X^+ n_X (1 - P_X(t)) - k_X^- P_X(t), \tag{6.9}$$

$$\frac{dP_Y(t)}{dt} = k_Y^+ n_Y (1 - P_Y(t)) - k_Y^- P_Y(t). \tag{6.10}$$

We also expect from the independency of both binding sites that the probabilities $P_{ij}(t)$ are given in terms of $P_X(t)$ and $P_Y(t)$ as follows:

$$P_{00}(t) = (1 - P_X(t))(1 - P_Y(t)), \tag{6.11}$$

$$P_{X0}(t) = P_X(t)(1 - P_Y(t)), \tag{6.12}$$

$$P_{0Y}(t) = (1 - P_X(t))P_Y(t), \tag{6.13}$$

$$P_{XY}(t) = P_X(t)P_Y(t). \tag{6.14}$$

To prove that these assumptions are correct, differentiate Eqs. (6.11)–(6.14) and substitute Eqs. (6.9) and (6.10) to recover Eqs. (6.5)–(6.8).

Equations (6.9) and (6.10) are not new to us since we encountered them in Chap. 3. The readers are encouraged to demonstrate that their general solutions are:

$$P_X(t) = \frac{n_X}{K_X + n_X} + \left(P_X^O - \frac{n_X}{K_X + n_X} \right) e^{-(k_X^+ n_X + k_X^-)t}, \tag{6.15}$$

$$P_Y(t) = \frac{n_Y}{K_Y + n_Y} + \left(P_Y^O - \frac{n_Y}{K_Y + n_Y} \right) e^{-(k_Y^+ n_Y + k_Y^-)t}, \tag{6.16}$$

with P_X^O and P_Y^O the initial values for $P_X(t)$ and $P_Y(t)$, $K_X = k_X^-/k_X^+$, and $K_X = k_X^-/k_X^+$. Equations (6.15) and (6.16) further imply that

$$\lim_{t\to\infty} P_X(t) = \frac{n_X}{K_X + n_X}, \quad \lim_{t\to\infty} P_Y(t) = \frac{n_Y}{K_Y + n_Y}, \tag{6.17}$$

and that the rates of convergence are, respectively, determined by $k_X^+ n_X + k_X^-$ and $k_Y^+ n_Y + k_Y^-$.

The expressions for $P_{ij}(t)$ can be computed by substituting Eqs. (6.15) and (6.16) into Eqs. (6.11)–(6.14), but the resulting formulas are too long to be informative. Let us, instead, calculate the $P_{ij}(t)$ stationary values from Eqs. (6.17):

$$\overline{P}_{00} = \frac{1}{\left(1 + \frac{n_X}{K_X}\right)\left(1 + \frac{n_Y}{K_Y}\right)}, \tag{6.18}$$

$$\overline{P}_{X0} = \frac{\frac{n_X}{K_X}}{\left(1 + \frac{n_X}{K_X}\right)\left(1 + \frac{n_Y}{K_Y}\right)}, \tag{6.19}$$

$$\overline{P}_{0Y} = \frac{\frac{n_Y}{K_Y}}{\left(1 + \frac{n_X}{K_X}\right)\left(1 + \frac{n_Y}{K_Y}\right)}, \tag{6.20}$$

$$\overline{P}_{XY} = \frac{\frac{n_X}{K_X}\frac{n_Y}{K_Y}}{\left(1 + \frac{n_X}{K_X}\right)\left(1 + \frac{n_Y}{K_Y}\right)}. \tag{6.21}$$

Given that the rate of convergence to the steady state of a complex system composed of two or more coupled subsystems is determined by the slowest subsystem (Strogatz 1994), the speed with which the probabilities $P_{ij}(t)$ converge to their stationary values is determined by the minimum value of $k_X^+ n_X + k_X^-$ and $k_Y^+ n_Y + k_Y^-$. When either n_X or n_Y are equal to zero, Eqs. (6.18)–(6.21) reduce to the well-known Michaelis–Menten equation that we found in Chap. 5 while studying the enzyme–substrate interaction.

6.3 Multiple Receptor Molecules

Let us now generalize our analysis to the case of n_T identical receptors. Assume that each receptor molecule has two specific binding sites for ligands X and Y, and that n_X and n_Y are kept constant. As before, the first step is to write down the corresponding master equation. Let $P(n_R, n_{RX}, n_{RY}; t)$ denote the probability that, at time t, there are n_R free receptor molecules, n_{RX} receptor molecules bound by a ligand X, and n_{RY} receptor molecules bound by a ligand Y. Since we are assuming a constant total number of receptors, the count of receptor molecules whose both binding sites are occupied is given by $n_{RXY} = n_T - n_R - n_{RX} - n_{RY}$. This is all the information we need to derive the master equation governing the dynamics of $P(n_R, n_{RX}, n_{RY}; t)$. Although quite long, it is worthwhile writing this equation in full for the sake of clarity:

$$
\begin{aligned}
\frac{dP(n_R, n_{RX}, n_{RY}; t)}{dt} &= k_X^+(n_R + 1)n_X P(n_R + 1, n_{RX} - 1, n_{RY}; t) \\
&- k_X^+ n_R n_X P(n_R, n_{RX}, n_{RY}; t) \\
&+ k_X^-(n_{RX} + 1) P(n_R - 1, n_{RX} + 1, n_{RY}; t) \\
&- k_X^- n_{RX} P(n_R, n_{RX}, n_{RY}; t) \\
&+ k_Y^+(n_R + 1)n_Y P(n_R + 1, n_{RX}, n_{RY} - 1; t) \\
&- k_Y^+ n_R n_Y P(n_R, n_{RX}, n_{RY}; t) \\
&+ k_Y^-(n_{RY} + 1) P(n_R - 1, n_{RX}, n_{RY} + 1; t) \\
&- k_Y^- n_{RY} P(n_R, n_{RX}, n_{RY}; t) \\
&+ k_Y^+(n_{RX} + 1)n_Y P(n_R, n_{RX} + 1, n_{RY}; t) \\
&- k_Y^+ n_{RX} n_Y P(n_R, n_{RX}, n_{RY}; t) \\
&+ k_Y^-(n_{RXY} + 1) P(n_R, n_{RX} - 1, n_{RY}; t) \\
&- k_Y^- n_{RXY} P(n_R, n_{RX}, n_{RY}; t) \\
&+ k_X^+(n_{RY} + 1)n_X P(n_R, n_{RX}, n_{RY} + 1; t)
\end{aligned}
$$

$$-k_X^+ n_{RY} n_X P(n_R, n_{RX}, n_{RY}; t)$$
$$+k_X^-(n_{RXY}+1) P(n_R, n_{RX}, n_{RY}-1; t)$$
$$-k_X^- n_{RXY} P(n_R, n_{RX}, n_{RY}; t) \qquad (6.22)$$

To understand Eq. (6.22) take a look at the reactions in Eqs. (6.1)–(6.4). The first two terms on the right-hand side of Eq. (6.22) correspond to the forward reaction in (6.1); the first term accounts for the case in which one of such reactions takes the system into the state (n_R, n_{RX}, n_{RY}), while the second term considers the reactions that take the system out of this state. In a similar way, the third and fourth terms on the right-hand side of Eq. (6.22) correspond to the backward reaction in (6.1), the fifth and sixth terms correspond to the forward reaction in (6.2), and so forth for the rest of the reactions.

The reader can imagine how difficult it is to solve Eq. (6.22) directly. However, one can construct its solution from Eqs. (6.11)–(6.14)—with $P_X(t)$ and $P_Y(t)$ as given by Eqs. (6.15) and (6.16)—and from the supposition that all the n_T receptor molecules are independent from each other. In summary, given that we know the probability distribution for all the states of a single molecule, the probability distribution for the n_T molecules is nothing but a multinomial distribution:

$$P(n_R, n_{RX}, n_{RY}; t) = \frac{n_T!}{n_R! n_{RX}! n_{RX}! n_{RXY}!} P_{00}^{n_R}(t) P_{X0}^{n_{RX}}(t) P_{0Y}^{n_{RY}}(t) P_{XY}^{n_{RXY}}(t), \qquad (6.23)$$

with $n_{RXY} = n_T - n_R - n_{RX} - n_{RY}$. The first term on the right-hand side of Eq. (6.23) accounts for the number of ways in which the n_T molecules can be arranged such that n_R of them are free, n_{RX} are bound by a ligand X, n_{RY} are bound by a ligand Y, and the rest are bound by both ligands. The rest of the terms on the right-hand side of Eq. (6.23) stands for the probability of each one of such configurations. Verifying that (6.23) is a solution to (6.22) is simple but tedious. However, the readers are encouraged to perform the demonstration because it is a very instructive exercise.

With the probability distribution it is possible to calculate the average number of molecules in each state:

$$N_i(t) = \sum_{n_i, n_{RX}, n_{RY}} n_i P(n_R, n_{RX}, n_{RY}), \quad i = R, RX, RY. \qquad (6.24)$$

At this point we make use of the multinomial distribution properties (Evans et al. 2000) to obtain

$$N_R(t) = n_T P_{00}(t), \qquad (6.25)$$
$$N_{RX}(t) = n_T P_{X0}(t), \qquad (6.26)$$
$$N_{RY}(t) = n_T P_{0Y}(t), \qquad (6.27)$$
$$N_{RXX}(t) = n_T P_{XY}(t), \qquad (6.28)$$

Finally, since the probabilities $P_{ij}(t)$ converge to the steady-state values given by (6.18)–(6.21) as $t \rightarrow \infty$, the average receptor counts in every state reach the following stationary values as time tends to infinity:

$$\overline{N}_R = n_T \frac{1}{\left(1 + \frac{n_X}{K_X}\right)\left(1 + \frac{n_Y}{K_Y}\right)}, \tag{6.29}$$

$$\overline{N}_{RX} = n_T \frac{\frac{n_X}{K_X}}{\left(1 + \frac{n_X}{K_X}\right)\left(1 + \frac{n_Y}{K_Y}\right)}, \tag{6.30}$$

$$\overline{N}_{RY} = n_T \frac{\frac{n_Y}{K_Y}}{\left(1 + \frac{n_X}{K_X}\right)\left(1 + \frac{n_Y}{K_Y}\right)}, \tag{6.31}$$

$$\overline{N}_{RXX} = n_T \frac{\frac{n_X}{K_X}\frac{n_Y}{K_Y}}{\left(1 + \frac{n_X}{K_X}\right)\left(1 + \frac{n_Y}{K_Y}\right)}. \tag{6.32}$$

Observe the similarity between Eqs. (6.18)–(6.21) and (6.29)–(6.32). In fact, what we have obtained are results that agree with the common notion of probability (Jaynes 2003): the average number of receptors found in a given state is given by the probability of such state, times the total number of receptors.

6.4 Deterministic Kinetic Analysis

To analyze the dynamics of the mean molecular counts, $N_i(t)$, differentiate Eq. (6.24) for all the i values and substitute Eq. (6.22) to obtain

$$\frac{dN_R(t)}{dt} = k_X^- N_{RX}(t) + k_Y^- N_{RY}(t) - (k_X^+ n_X + k_Y^+ n_Y) N_R(t), \tag{6.33}$$

$$\frac{dN_{RX}(t)}{dt} = k_X^+ n_X N_R(t) + k_Y^- N_{RXY}(t) - (k_X^- + k_Y^+ n_Y) N_{RX}(t), \tag{6.34}$$

$$\frac{dN_{RY}(t)}{dt} = k_Y^+ n_Y N_R(t) + k_X^- N_{RXY}(t) - (k_Y^- + k_X^+ n_X) N_{RY}(t), \tag{6.35}$$

with $N_{RXY}(t) = n_T - N_R(t) - N_{RX}(t) - N_{RY}(t) - N_{RXY}(t)$.

By solving the algebraic equations resulting from setting $dN_i/dt = 0$, one can compute the stationary states available for the system. The readers are encouraged to do it either manually or with the aid of a computer algebra system like `Mathematica` or `Maple`, and prove that the only existing stationary state is given by Eqs. (6.29)–(6.32), in agreement with the analysis in the previous section.

The system of differential equations (6.33)–(6.35) can also be solved without much trouble. To do it define

$$x = N_{RX} - \overline{N}_{RX}, \quad y = N_{RY} - \overline{N}_{RY}, \quad z = N_R - \overline{N}_R, \tag{6.36}$$

calculate the derivatives, and substitute Eqs. (6.33)–(6.35) to get

$$\frac{dx}{dt} = -(k_X^- + k_Y^+ n_Y + k_Y^-)x - k_Y^- y + (k_X^+ n_X - k_Y^-)z. \tag{6.37}$$

$$\frac{dy}{dt} = -k_X^- x - (k_Y^- + k_X^+ n_X + k_X^-)y + (k_Y^+ n_Y - k_X^-)z, \tag{6.38}$$

$$\frac{dz}{dt} = k_X^- x + k_Y^- y - (k_X^+ n_X + k_Y^+ n_Y)z, \tag{6.39}$$

This is a linear system that can be written in vectorial form as $\dot{\mathbf{x}} = \mathbf{A}\mathbf{x}$, with $\mathbf{x} = (x, y, z)^T$, and

$$\mathbf{A} = \begin{bmatrix} -(k_X^- + k_Y^+ n_Y + k_Y^-) & -k_Y^- & (k_X^+ n_X - k_Y^-) \\ -k_X^- & -(k_Y^- + k_X^+ n_X + k_X^-) & (k_Y^+ n_Y - k_X^-) \\ k_X^- & k_Y^- & -(k_X^+ n_X + k_Y^+ n_Y) \end{bmatrix}.$$

The general solution of the differential-equation system (6.37)–(6.39) is (Zill 2008):

$$\mathbf{x}(t) = \sum_{i=1}^{3} C_i \mathbf{v}_i e^{\lambda_i t}, \tag{6.40}$$

with $\mathbf{v}_1, \mathbf{v}_2, \mathbf{v}_3$ the eigenvectors of matrix $\mathbf{A}$ and $\lambda_1, \lambda_2, \lambda_3$ the corresponding eigenvalues. After performing the corresponding calculations, the eigenvector of $\mathbf{A}$ are:

$$\mathbf{v}_1 = \begin{pmatrix} -1 \\ k_Y^+/k_Y^- \\ 1 \end{pmatrix}, \quad \mathbf{v}_2 = \begin{pmatrix} k_X^+/k_X^- \\ -1 \\ 1 \end{pmatrix}, \quad \mathbf{v}_3 = \begin{pmatrix} -1 \\ -1 \\ 1 \end{pmatrix}, \tag{6.41}$$

while the corresponding eigenvalues are

$$\lambda_1 = -(k_X^+ + k_X^-), \; \lambda_2 = -(k_Y^+ + k_Y^-), \; \lambda_3 = -(k_X^+ + k_X^- + k_Y^+ + k_Y^-). \tag{6.42}$$

Assume that following initial conditions: $x(0) = x_0$, $y(0) = x_0$, and $y(0) = y_0$. It then follows from Eq. (6.40) that

$$(x_0, y_0, z_0)^T = C_1\mathbf{v}_1 + C_2\mathbf{v}_2 + C_3\mathbf{v}_3.$$

After substituting Eq. (6.41) we get

$$x_0 = -C_1 + C_2/K_X - C_3, \; y_0 = C_1/K_Y - C_2 - C_3, \; z_0 = C_1 + C_2 + C_3, \tag{6.43}$$

with $K_X = k_X^-/k_X^+$ and $K_Y = k_Y^-/k_Y^+$. By solving for C_1, C_2, and C_3 from the equation system in (6.43), we obtain the following expressions for these originally undetermined constants in terms of the initial condition $\mathbf{X}(0) = \mathbf{x}_0$:

$$C_1 = (y_0 + y_0)\frac{K_Y}{1 + K_Y}, \tag{6.44}$$

$$C_2 = (x_0 + y_0)\frac{K_X}{1 + K_X}, \tag{6.45}$$

$$C_3 = z_0\frac{1 - K_X K_Y}{(1 + K_X)(1 + K_Y)} - x_0\frac{K_X}{1 + K_X} - y_0\frac{K_Y}{1 + K_Y}. \tag{6.46}$$

From the above results, the solutions to the differential-equation system (6.37)–(6.39) are

$$x(t) = -C_1 e^{\lambda_1 t} + \frac{C_2}{K_X} e^{\lambda_2 t} - C_3 e^{\lambda_3 t},$$

$$y(t) = \frac{C_1}{K_Y} e^{\lambda_1 t} - C_2 e^{\lambda_2 t} - C_3 e^{\lambda_3 t},$$

$$z(t) = C_1 e^{\lambda_1 t} + C_2 e^{\lambda_2 t} + C_3 e^{\lambda_3 t},$$

with C_1, C_2, and C_3 as given in Eqs. (6.44)–(6.46).

Finally, if we revert the change of variables in Eq. (6.36), the solutions to the system of ordinary differential equations in (6.33)–(6.35) result to be

$$N_{RX}(t) = \overline{N}_{RX} - C_1 e^{\lambda_1 t} + \frac{C_2}{K_X} e^{\lambda_2 t} - C_3 e^{\lambda_3 t}, \tag{6.47}$$

$$N_{RY}(t) = \overline{N}_{RY} + \frac{C_1}{K_Y} e^{\lambda_1 t} - C_2 e^{\lambda_2 t} - C_3 e^{\lambda_3 t}, \tag{6.48}$$

$$N_R(t) = \overline{N}_R + C_1 e^{\lambda_1 t} + C_2 e^{\lambda_2 t} + C_3 e^{\lambda_3 t}, \tag{6.49}$$

where C_1, C_2, and C_3 are given by Eqs. (6.44)–(6.46), while λ_1, λ_2, and λ_3 are as in Eq. (6.47). Notice that in the calculation of the constants C_i we have to make the substitution

$$x_0 = N_{RX}^0 - \overline{N}_{RX}, \quad y_0 = N_{RY}^0 - \overline{N}_{RY}, \quad z_0 = N_R^0 - \overline{N}_R,$$

where $N_{RX}^0 = N_{RX}(0)$, $N_{RY}^0 = N_{RY}(0)$, and $N_R^0 = N_R(0)$.

Two important conclusions follow from Eqs. (6.47)–(6.49):

- Since all the system eigenvalues are negative (6.47), $N_{RX}(t)$, $N_{RY}(t)$, and $N_R(t)$ converge to their respective stationary values as $t \to \infty$.

- The rate of convergence is determined by the eigenvalue with the smallest absolute value (Strogatz 1994). In this case, it has to be either $|\lambda_1| = k_X^+ + k_X^-$ or $|\lambda_2| = k_Y^+ + k_Y^-$, because $|\lambda_3| = |\lambda_1| + |\lambda_2|$. In other words, the rate of convergence to the stationary state is determined by the slowest binding–unbinding reaction set.

To end this section it is important to recall that Eqs. (6.47)–(6.49) describe the temporal behavior of the average molecular counts for n_{RX}, n_{RY}, and n_R. So, the question arises of how good a description of these equations are for a single experiment or a single simulation carried out using Gillespie's algorithm. We expect that the temporal evolution of an individual stochastic simulation will present fluctuations around the corresponding average values. Moreover, the results in the previous section predict that if we run several simulations, count at a given time the number of molecules in each state, and calculate the histograms, they will correspond to the distributions derived from the multinomial distribution in (6.23). I leave for the readers the exercise of calculating the probability distributions $P(n_{RX})$, $P(n_{RY})$, and $P(n_R)$ out of Eq. (6.23), as well as the corresponding mean values and standard deviations. The results of these calculations shall show that, in all cases, the coefficients of variation are proportional to $1/\sqrt{n_T}$. Therefore, the larger the total number of receptors, the better the deterministic description given by Eqs. (6.47)–(6.49). This also means that, at least for one of the cases of interest given at the beginning of the chapter (gene regulation), the deterministic description is not good enough because the number of gene copies within one cell is generally one or two.

6.5 Thermodynamic Analysis

As in previous examples, we will make use of an energy landscape to analyze the reaction scheme in (6.1)–(6.4) from a thermodynamic perspective and make the connection with the system dynamics. Since there are two possible paths to go from $R + X + Y$ to R_{XY}, the complete energy landscape is three dimensional; and this makes it difficult to visualize. Fortunately, we do not need the complete picture. It is enough to know how the energy changes along the two possible trajectories: those corresponding to reactions (6.1) and (6.3), and to reactions (6.2) and (6.4). A schematic representation of such profiles is presented in Fig. 6.2.

From Kramer's theory (Van Kampen 1992) and following the same argumentation that lead to Eqs. (3.29) and (3.30), the reaction rates in (6.1)–(6.4) can be written in terms of the various energy levels in the energy profiles of Fig. 6.2 as

$$k_X^+ = \beta e^{\mu_R^O/k_BT} e^{\mu_X^O/k_BT} e^{\mu_Y^O/k_BT} e^{-\xi_1/k_BT} \tag{6.50}$$

$$= \beta e^{\mu_{RY}^O/k_BT} e^{\mu_X^O/k_BT} e^{-\xi_2/k_BT}, \tag{6.51}$$

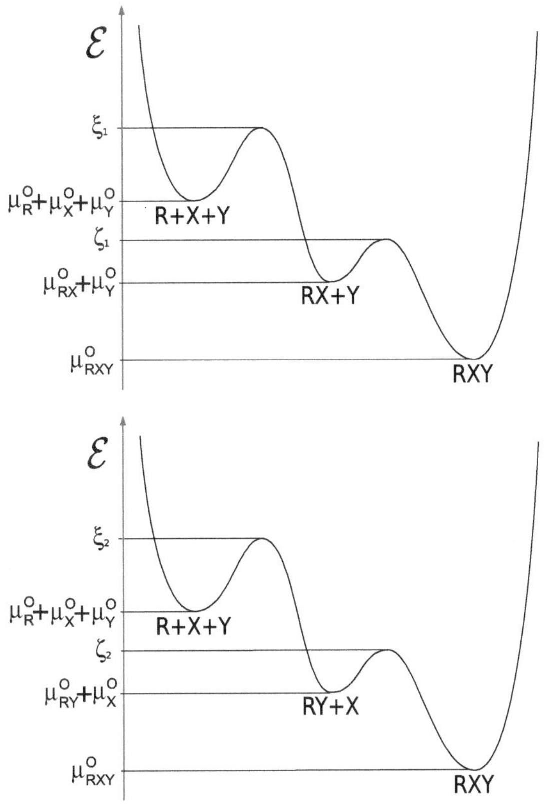

Fig. 6.2 *Top*: schematic representation of the energy profile along the reactions (6.1) and (6.3). *Top*: schematic representation of the energy profile along the reactions (6.2) and (6.4)

$$k_X^- = \beta e^{\mu_{RX}^O/k_BT} e^{\mu_Y^O/k_BT} e^{-\xi_1/k_BT} \tag{6.52}$$

$$= \beta e^{\mu_{RXY}^O/k_BT} e^{-\zeta_2/k_BT}, \tag{6.53}$$

$$k_Y^+ = \beta e^{\mu_R^O/k_BT} e^{\mu_X^O/k_BT} e^{\mu_Y^O/k_BT} e^{-\xi_2/k_BT} \tag{6.54}$$

$$= \beta e^{\mu_{RX}^O/k_BT} e^{\mu_Y^O/k_BT} e^{-\zeta_1/k_BT}, \tag{6.55}$$

$$k_Y^- = \beta e^{\mu_{RY}^O/k_BT} e^{\mu_X^O/k_BT} e^{-\xi_1/k_BT} \tag{6.56}$$

$$= \beta e^{\mu_{RXY}^O/k_BT} e^{-\zeta_2/k_BT}. \tag{6.57}$$

Take the logarithm in Eqs. (6.50) and (6.51), multiply by k_BT, and solve for $\xi_1 - \zeta_2$. Do the same with Eqs. (6.52) and (6.53), equate with the result of the previous operations, and solve for μ_{RXY}^O to obtain

$$\mu_{RXY}^O = \mu_{RX}^O + \mu_{RY}^O - \mu_R^O. \tag{6.58}$$

Let us save this result for the time being. Later on it will be useful and we shall interpret it. Just have in mind that it comes from the fact that the forward and backward rates of reactions (6.1) and (6.4) are equal. The same result can be obtained by making the equivalent algebraic operations with Eqs. (6.4)–(6.7). I leave the proof of this assertion as an exercise for the readers.

Observe that both energy profiles look like the ones previously studied in Chap. 4 (see Fig. 4.4). In Sect. 4.4 we saw that when there is a constant number of molecules jumping between the states corresponding to the three minima of the energy landscape, the stationary state corresponds to thermodynamic equilibrium. Moreover, thermodynamic equilibrium is characterized by equal chemical potentials in all three states. In the present case we have a similar situation: a constant number of receptors are switching between different binding states. Thus, when the stationary condition is imposed to both reaction pathways we get

$$\mu_R + \mu_X + \mu_Y = \mu_{RX} + \mu_Y = \mu_{RXY}, \quad \mu_R + \mu_X + \mu_Y = \mu_{RY} + \mu_X = \mu_{RXY}.$$

Or simply:

$$\mu_R + \mu_X + \mu_Y = \mu_{RX} + \mu_Y = \mu_{RY} + \mu_X = \mu_{RXY}. \tag{6.59}$$

Recall that $\mu_i = \mu_i^O + k_B T \ln \overline{N}_i$. Hence, by substituting into Eq. (6.59) we obtain

$$\begin{aligned} \overline{N}_R e^{\mu_R^O/k_B T} n_X e^{\mu_X^O/k_B T} n_Y e^{\mu_Y^O/k_B T} &= \overline{N}_{RX} e^{\mu_{RX}^O/k_B T} n_Y e^{\mu_Y^O/k_B T} \\ &= \overline{N}_{RY} e^{\mu_{RY}^O/k_B T} n_X e^{\mu_X^O/k_B T} \\ &= \overline{N}_{RXY} e^{\mu_{RXY}^O/k_B T}. \end{aligned} \tag{6.60}$$

Furthermore, by solving for $\overline{N}_{RX}$, $\overline{N}_{RY}$, and $\overline{N}_{RXY}$ in terms of $\overline{N}_R$ we get

$$\overline{N}_{RX} = \overline{N}_R \frac{n_X}{e^{(\mu_{RX}^O - \mu_R^O - \mu_X^O)/k_B T}}, \tag{6.61}$$

$$\overline{N}_{RY} = \overline{N}_R \frac{n_Y}{e^{(\mu_{RY}^O - \mu_R^O - \mu_Y^O)/k_B T}}, \tag{6.62}$$

$$\overline{N}_{RXY} = \overline{N}_R \frac{n_X n_Y}{e^{(\mu_{RXY}^O - \mu_R^O - \mu_X^O - \mu_Y^O)/k_B T}}, \tag{6.63}$$

Notice that the term $\mu_{RX}^O - \mu_R^O - \mu_X^O$ in the exponential of the right-hand side of Eq. (6.60) corresponds to the energy difference between states $R + X + Y$ and $R_X + Y$. Similarly, the term $\mu_{RY}^O - \mu_R^O - \mu_Y^O$ in Eq. (6.60) is the energy difference between states $R + X + Y$ and $R_Y + X$, while $\mu_{RXY}^O - \mu_R^O - \mu_X^O - \mu_Y^O$ is the energy difference between states R_{XY} and $R + X + Y$. However, if we substitute Eq. (6.58), we obtain

$$\mu_{RXY}^O - \mu_R^O - \mu_X^O - \mu_Y^O = (\mu_{RX}^O - \mu_R^O - \mu_X^O) + (\mu_{RY}^O - \mu_R^O - \mu_Y^O). \tag{6.64}$$

That is, the energy difference between states R_{XY} and $R + X + Y$ is equal to the sum of the energy differences between states $R_X + Y$ and $R + X + Y$, and between states $R_Y + X$ and $R + X + Y$. Furthermore, given that Eq. (6.58) is a direct consequence of the assumption that the two different binding sites in the receptor molecule are independent, we have that this independence is equivalent to the additivity of the energy differences. We shall return to this point later.

Substitute Eq. (6.64) into Eq. (6.63) to obtain

$$\overline{N}_{RXY} = \overline{N}_R \frac{n_X}{e^{(\mu_{RX}^O - \mu_R^O - \mu_X^O)/k_B T}} \frac{n_Y}{e^{(\mu_{RY}^O - \mu_R^O - \mu_Y^O)/k_B T}} \tag{6.65}$$

From Eqs. (6.61), (6.62), and (6.65), plus the assumption that the total number of receptors is constant:

$$n_T = \overline{N}_R + \overline{N}_{RX} + \overline{N}_{RY} + \overline{N}_{RXY}$$

we get the following expressions for $\overline{N}_R$, $\overline{N}_{RX}$,$\overline{N}_{RY}$, and $\overline{N}_{RXY}$:

$$\overline{N}_R = n_T \frac{1}{\left(1 + n_X e^{-(\mu_{RX}^O - \mu_R^O - \mu_X^O)/k_B T}\right)\left(1 + n_Y e^{-(\mu_{RY}^O - \mu_R^O - \mu_Y^O)/k_B T}\right)}, \tag{6.66}$$

$$\overline{N}_{RX} = n_T \frac{n_X e^{-(\mu_{RX}^O - \mu_R^O - \mu_X^O)/k_B T}}{\left(1 + n_X e^{-(\mu_{RX}^O - \mu_R^O - \mu_X^O)/k_B T}\right)\left(1 + n_Y e^{-(\mu_{RY}^O - \mu_R^O - \mu_Y^O)/k_B T}\right)}, \tag{6.67}$$

$$\overline{N}_{RY} = n_T \frac{n_Y e^{-(\mu_{RY}^O - \mu_R^O - \mu_Y^O)/k_B T}}{\left(1 + n_X e^{-(\mu_{RX}^O - \mu_R^O - \mu_X^O)/k_B T}\right)\left(1 + n_Y e^{-(\mu_{RY}^O - \mu_R^O - \mu_Y^O)/k_B T}\right)}, \tag{6.68}$$

$$\overline{N}_{RXY} = n_T \frac{n_X e^{-(\mu_{RX}^O - \mu_R^O - \mu_X^O)/k_B T} n_Y e^{-(\mu_{RY}^O - \mu_R^O - \mu_Y^O)/k_B T}}{\left(1 + n_X e^{-(\mu_{RX}^O - \mu_R^O - \mu_X^O)/k_B T}\right)\left(1 + n_Y e^{-(\mu_{RY}^O - \mu_R^O - \mu_Y^O)/k_B T}\right)}. \tag{6.69}$$

Let us make a pause at this point to introduce some useful thermodynamic concepts to further analyze the results expressed in Eqs. (6.66)–(6.69). We know from Eq. (2.6) that, except for an additive constant, the Gibbs free energy of a chemical system kept at constant pressure and temperature can be expressed as

$$G = \sum_i \mu_i \overline{N}_i.$$

If we substitute $\mu_i = \mu_i^O + k_B T \ln \overline{N}_i$ the above equation transforms into

$$G = \sum_i (\mu_i^O + k_B T \ln \overline{N}_i) \overline{N}_i = G^O + k_B T \sum_i \overline{N}_i \ln \overline{N}_i,$$

where

$$G^O = \sum_i \mu_i^O \overline{N}_i.$$

Assume that there is a chemical reaction taking place in the system. We can define the free energy change of this chemical reaction as usual (de Groot and Mazur 2013):

$$\Delta G = \frac{dG}{d\xi} = \frac{dG^O}{d\xi} + \frac{d}{d\xi}\sum_i \overline{N}_i \ln \overline{N}_i = \Delta G^O + \frac{d}{d\xi}\sum_i \overline{N}_i \ln \overline{N}_i,$$

in which

$$\Delta G^O = \sum_i \mu_i^O \frac{d\overline{N}_i}{d\xi}$$

is called the standard free energy change (or the free energy change under standard conditions). Variable ξ is usually called the degree of advance of the reaction, and measures the net number of forward individual chemical reactions that have occurred from the start of the experiment. Let us now focus on the terms $d\overline{N}_i/d\xi$. They can be interpreted as the amount of change of the molecular count of species i per unitary increment of the reaction degree of advancement. In other words, they measure how many new molecules appear (of disappear if the derivative is negative) per unitary forward reaction. For this reason they are called the stoichiometric coefficients. Let us denote them as $\nu_i = d\overline{N}_i/d\xi$. Hence

$$\Delta G^O = \sum_i \mu_i^O \nu_i. \tag{6.70}$$

Upon applying the definition in (6.70) we obtain

$$\Delta G_X^O = \mu_{RX}^O - \mu_R^O - \mu_X^O, \tag{6.71}$$

$$\Delta G_Y^O = \mu_{RY}^O - \mu_R^O - \mu_Y^O. \tag{6.72}$$

If the binding of the two ligands to a single receptor is viewed as a single global reaction (rather than two sequential reactions), we can also define

$$\Delta G_{XY}^O = \mu_{RXY}^O - \mu_R^O - \mu_X^O - \mu_Y^O. \tag{6.73}$$

The result in (6.64) guaranties that

$$\Delta G_{XY}^O = \Delta G_X^O + \Delta G_Y^O.$$

One way to interpret this last result is to choose the zero energy level in the energy landscape of Fig. 6.2 in such a way that $\mu_R^O + \mu_X^O + \mu_Y^0 = 0$. With this, $\Delta G_X^O, \Delta G_Y^O, \Delta G_{XY}^O < 0$ can be understood as the binding energies of states $R_X + Y$, $R_Y + X$, and R_{XY}, respectively. Thus, when the two ligands are bound to the receptor, the binding energy of the complex results to be the addition of the individual binding energies associated with each ligand. It is important to remark that this additivity property is a consequence of the independency of the X and Y binding sites.

To continue with our analysis substitute Eqs. (6.71) and (6.72) into Eqs. (6.66)–(6.69) to obtain:

$$\overline{N}_R = n_T \frac{1}{\left(1 + n_X e^{-\Delta G_X^O/k_B T}\right)\left(1 + n_Y e^{-\Delta G_Y^O/k_B T}\right)}, \tag{6.74}$$

$$\overline{N}_{RX} = n_T \frac{n_X e^{-\Delta G_X^O/k_B T}}{\left(1 + n_X e^{-\Delta G_X^O/k_B T}\right)\left(1 + n_Y e^{-\Delta G_Y^O/k_B T}\right)}, \tag{6.75}$$

$$\overline{N}_{RY} = n_T \frac{n_Y e^{-\Delta G_Y^O/k_B T}}{\left(1 + n_X e^{-\Delta G_X^O/k_B T}\right)\left(1 + n_Y e^{-\Delta G_Y^O/k_B T}\right)}, \tag{6.76}$$

$$\overline{N}_{RXY} = n_T \frac{n_X e^{-\Delta G_X^O/k_B T} n_Y e^{-\Delta G_Y^O/k_B T}}{\left(1 + n_X e^{-\Delta G_X^O/k_B T}\right)\left(1 + n_Y e^{-\Delta G_Y^O/k_B T}\right)}. \tag{6.77}$$

It is not difficult to prove that Eqs. (6.50)–(6.57), together with Eqs. (6.71)–(6.72) imply that

$$e^{-\Delta G_X^O/k_B T} = k_X^+/k_X^- = 1/K_X, \quad e^{-\Delta G_Y^O/k_B T} = k_Y^+/k_Y^- = 1/K_y, \tag{6.78}$$

in which K_X and K_Y are known as the dissociation constants of their respective reactions. Finally, by substitution (6.78) into Eqs. (6.74)–(6.77) we get

$$\overline{N}_R = n_T \frac{1}{\left(1 + \frac{n_X}{K_X}\right)\left(1 + \frac{n_Y}{K_Y}\right)},$$

$$\overline{N}_{RX} = n_T \frac{\frac{n_X}{K_X}}{\left(1 + \frac{n_X}{K_X}\right)\left(1 + \frac{n_Y}{K_Y}\right)},$$

$$\overline{N}_{RY} = n_T \frac{\frac{n_Y}{K_Y}}{\left(1 + \frac{n_X}{K_X}\right)\left(1 + \frac{n_Y}{K_Y}\right)},$$

$$\overline{N}_{RXX} = n_T \frac{\frac{n_X}{K_X}\frac{n_Y}{K_Y}}{\left(1 + \frac{n_X}{K_X}\right)\left(1 + \frac{n_Y}{K_Y}\right)}.$$

That is, we have recovered Eqs. (6.29)–(6.32). Recall that we previously obtained these equations from the stochastic (master equation) and deterministic (chemical kinetics) approaches, confirming once more the unity of the different existing formulations to studying the dynamics of chemical-reaction systems.

6.6 Summary

In the present chapter we analyzed the dynamical and thermodynamical behavior of a ubiquitous process in biology: the binding of one or more ligands to one receptor. We studied in detail the case when a single receptor molecule is present, as well as when there are several receptors. Not only we were able to prove once more the unity of the various approaches previously introduced, but also derived some interesting conclusions. For instance, we confirmed that the chemical kinetics equations govern the evolutions of the average molecular counts, as computed from the master equation approach. We also proved that, in the present system, the stationary state is compatible with chemical and thermodynamic equilibria, and showed that the stationary state is unique and stable.

Chapter 7
Cooperativity

Abstract In the present chapter we generalize the results in the previous chapter to study one of the most fascinating biochemical phenomena: the cooperative interaction between two or more ligands that bind a single receptor (or simply cooperativity). Once more, we make use of our previous knowledge to carefully analyze the dynamic and thermodynamic characteristics of this phenomenon.

7.1 The Importance of Cooperativity

Cooperativity is a phenomenon displayed by enzymes or receptors that have multiple binding sites whose affinity for a ligand is increased, positive cooperativity, or decreased, negative cooperativity, upon the binding of a ligand to a neighboring binding site. One of the most striking examples of cooperativity occurs in hemoglobin. The affinity of this molecule's four binding sites for oxygen is increased above that of the unbound hemoglobin when the first oxygen molecule binds. This behavior is essential to explain the efficiency of this molecule to transport and exchange oxygen and carbon dioxide. Furthermore, as we shall see, cooperativity is a very good example of how the dynamical and thermodynamical perspectives nicely complement each other. By putting them together we can get a better and more profound understanding of this biochemical phenomenon.

7.2 One Receptor-Two Ligands, Stochastic Kinetic Analysis

Consider once more a single receptor molecule with two binding sites, one specific for X ligands and the other specific for Y ligands. We assume that the processes through which the receptor's binding sites are occupied and unoccupied are as depicted in Fig. 6.1. However, contrarily to the analysis in Chap. 6, we do not assume in here that the binding-unbinding processes in one site are independent from those in the other. Instead, we suppose that the probability that a ligand binds to its specific

M. Santillán, *Chemical Kinetics, Stochastic Processes, and Irreversible Thermodynamics*, Lecture Notes on Mathematical Modelling in the Life Sciences,
DOI 10.1007/978-3-319-06689-9_7,

site is independent of the other site state. However, the probability that a given ligand detaches from its binding site decreases when the other site is occupied. In other words, once the two ligands are bound, they interact in such a way that the whole complex stability is increased. As previously discussed, this phenomenon is an instance of cooperativity. The above discussion can be summarized in the following reaction scheme

$$R + X \underset{k_X^- n_{RX}}{\overset{k_X^+ n_R n_X}{\rightleftharpoons}} R_X, \tag{7.1}$$

$$R + Y \underset{k_Y^- n_{RY}}{\overset{k_Y^+ n_R n_Y}{\rightleftharpoons}} R_Y, \tag{7.2}$$

$$R_X + Y \underset{k_Y^- n_{RXY}/k_C}{\overset{k_Y^+ n_{RX} n_Y}{\rightleftharpoons}} R_{XY}, \tag{7.3}$$

$$R_Y + X \underset{k_X^- n_{RXY}/k_C}{\overset{k_X^+ n_{RY} n_X}{\rightleftharpoons}} R_{XY}. \tag{7.4}$$

As before, k_X^+ and k_X^- respectively represent the propensities (probability per unit time) for the binding and unbinding (when site Y is empty) of a molecule X to its corresponding site; k_Y^+ and k_Y^- are the corresponding propensities for the binding and unbinding (when site X is empty) of a molecule Y; n_X and n_Y are the molecular counts of ligands X and Y, respectively; while n_R, n_{RX}, n_{RY}, and n_{RXY} correspond to the number of receptor molecules that are free, bound by a ligand X, bound by a ligand Y, and bound by both ligands, respectively. Parameter k_C is a constant whose value measures how strong cooperativity is. $k_C > 1$ means positive cooperativity. On the contrary, $k_C < 1$ denotes negative cooperativity, meaning that the stability of the R_{XY} complex decreases due to the interaction between the X and Y ligands.

If we consider a single receptor molecule that, due to the binding and unbinding of ligands X and Y, switches between its four available states: $(0, 0)$, $(X, 0)$, $(0, Y)$, and (X, Y), the master-equation family governing its dynamics is:

$$\frac{dP_{00}(t)}{dt} = k_X^- P_{X0}(t) - k_X^+ n_X P_{00}(t) + k_Y^- P_{0Y}(t) - k_Y^+ n_Y P_{00}(t), \tag{7.5}$$

$$\frac{dP_{X0}(t)}{dt} = k_X^+ n_X P_{00}(t) - k_X^- P_{X0}(t) + k_Y^-/k_C P_{XY}(t) - k_Y^+ n_Y P_{X0}(t), \tag{7.6}$$

$$\frac{dP_{0Y}(t)}{dt} = k_Y^+ n_X P_{00}(t) - k_Y^- P_{0Y}(t) + k_X^-/k_C P_{XY}(t) - k_X^+ n_Y P_{0Y}(t), \tag{7.7}$$

$$\frac{dP_{XY}(t)}{dt} = k_Y^+ n_Y P_{X0}(t) - k_Y^-/k_C P_{XY}(t) + k_X^+ n_X P_{0Y}(t) - k_X^-/k_C P_{XY}(t), \tag{7.8}$$

where $P_{ij}(t)$ is the probability that the receptor molecule is in state (i, j) at time t, while n_X and n_Y correspond to the X and Y molecule counts. Solving the differential equations in (7.5)–(7.8) is not as easy as it was when both binding sites were independent. However, in many cases the stationary solutions are highly informative. The reader won't find much difficulty to demonstrate that

$$\overline{P}_{00} = \frac{1}{1 + \frac{n_X}{K_X} + \frac{n_Y}{K_Y} + k_C \frac{n_X}{K_X}\frac{n_Y}{K_Y}}, \tag{7.9}$$

$$\overline{P}_{X0} = \frac{\frac{n_X}{K_X}}{1 + \frac{n_X}{K_X} + \frac{n_Y}{K_Y} + k_C \frac{n_X}{K_X}\frac{n_Y}{K_Y}}, \tag{7.10}$$

$$\overline{P}_{0Y} = \frac{\frac{n_Y}{K_Y}}{1 + \frac{n_X}{K_X} + \frac{n_Y}{K_Y} + k_C \frac{n_X}{K_X}\frac{n_Y}{K_Y}}, \tag{7.11}$$

$$\overline{P}_{XY} = \frac{k_C \frac{n_X}{K_X}\frac{n_Y}{K_Y}}{1 + \frac{n_X}{K_X} + \frac{n_Y}{K_Y} + k_C \frac{n_X}{K_X}\frac{n_Y}{K_Y}}, \tag{7.12}$$

constitute a stationary solution for the equation system (7.5)–(7.8).

A comparison between Eqs. (7.9)–(7.12) and Eqs. (6.18)–(6.21) reveals that cooperativity changes the relative probability of state (X, Y). Positive cooperativity ($k_C > 1$) increases the stationary probability of state (X, Y). This agrees with the fact that the propensity for the detachment of a ligand (either X or Y) decreases in proportion to the value of k_C. Conversely, the relative probability of state (X, Y) decreases in case of negative cooperativity ($k_C < 1$).

Consider now the case in which n_T independent receptor molecules interact with a constant number n_X of X ligands, and a constant number n_Y of Y ligands. Following the procedure leading to Eq. (6.22) it is straightforward to derive the following master equation for such a system:

$$\begin{aligned}
\frac{dP(n_R, n_{RX}, n_{RY}; t)}{dt} &= k_X^+ (n_R + 1) n_X P(n_R + 1, n_{RX} - 1, n_{RY}; t) \\
&- k_X^+ n_R n_X P(n_R, n_{RX}, n_{RY}; t) \\
&+ k_X^- (n_{RX} + 1) P(n_R - 1, n_{RX} + 1, n_{RY}; t) \\
&- k_X^- n_{RX} P(n_R, n_{RX}, n_{RY}; t) \\
&+ k_Y^+ (n_R + 1) n_Y P(n_R + 1, n_{RX}, n_{RY} - 1; t) \\
&- k_Y^+ n_R n_Y P(n_R, n_{RX}, n_{RY}; t) \\
&+ k_Y^- (n_{RY} + 1) P(n_R - 1, n_{RX}, n_{RY} + 1; t) \\
&- k_Y^- n_{RY} P(n_R, n_{RX}, n_{RY}; t) \\
&+ k_Y^+ (n_{RX} + 1) n_Y P(n_R, n_{RX} + 1, n_{RY}; t)
\end{aligned}$$

$$
\begin{aligned}
&-k_Y^+ n_{RX} n_Y P(n_R, n_{RX}, n_{RY}; t) \\
&+k_Y^- / k_C \cdot (n_{RXY} + 1) P(n_R, n_{RX} - 1, n_{RY}; t) \\
&-k_Y^- / k_C \cdot n_{RXY} P(n_R, n_{RX}, n_{RY}; t) \\
&+k_X^+ (n_{RY} + 1) n_X P(n_R, n_{RX}, n_{RY} + 1; t) \\
&-k_X^+ n_{RY} n_X P(n_R, n_{RX}, n_{RY}; t) \\
&+k_X^- / k_C \cdot (n_{RXY} + 1) P(n_R, n_{RX}, n_{RY} - 1; t) \\
&-k_X^- / k_C \cdot n_{RXY} P(n_R, n_{RX}, n_{RY}; t)
\end{aligned}
\tag{7.13}
$$

Thanks to the independence of the n_T receptor molecules, the solution to Eq. (7.13) can be constructed from the solutions to Eqs. (7.9)–(7.12). If $P_{00}(t)$, $P_{X0}(t)$, $P_{0Y}(t)$, and $P_{XY}(t)$ are the solutions to Eqs. (7.9)–(7.12), then the solution $P(n_R, n_{RX}, n_{RY}; t)$ to Eq. (7.13) is

$$
P(n_R, n_{RX}, n_{RY}; t) = \frac{n_T!}{n_R! n_{RX}! n_{RY}! n_{RXY}!} P_{00}^{n_R}(t) P_{X0}^{n_{RX}}(t) P_{0Y}^{n_{RY}}(t) P_{XY}^{n_{RXY}}(t).
\tag{7.14}
$$

This can be tested by differentiating Eq. (7.14), substituting Eqs. (7.9)–(7.12), and proving that the obtained result is equal to the one obtained after substituting Eq. (7.14) into the right-hand side of Eq. (7.13).

Equation (7.14) further implies that the master equation in (7.13) has a stationary solution given by the multinomial probability distribution:

$$
\overline{P}(n_R, n_{RX}, n_{RY}) = \frac{n_T!}{n_R! n_{RX}! n_{RY}! n_{RXY}!} \overline{P}_{00}^{n_R} \overline{P}_{X0}^{n_{RX}} \overline{P}_{0Y}^{n_{RY}} \overline{P}_{XY}^{n_{RXY}},
\tag{7.15}
$$

with $\overline{P}_{00}$, $\overline{P}_{X0}$, $\overline{P}_{0Y}$, and $\overline{P}_{XY}$ as given by Eqs. (7.9)–(7.12). From the properties of the multinomial distribution (Evans et al. 2000), it is possible to compute the means and the standard deviations for $n_R(t)$, $n_{RX}(t)$, $n_{RY}(t)$, and $n_{RXY}(t)$ as follows

$$
N_R(t) = n_T P_{00}(t),
\tag{7.16}
$$

$$
N_{RX}(t) = n_T P_{X0}(t),
\tag{7.17}
$$

$$
N_{RY}(t) = n_T P_{0Y}(t),
\tag{7.18}
$$

$$
N_{RXY}(t) = n_T P_{XY}(t).
\tag{7.19}
$$

and

$$
\sigma_R(t) = \sqrt{n_T P_{00}(t)[1 - P_{00}(t)]},
\tag{7.20}
$$

$$
\sigma_{RX}(t) = \sqrt{n_T P_{X0}(t)[1 - P_{X0}(t)]},
\tag{7.21}
$$

$$\sigma_{RY}(t) = \sqrt{n_T P_{0Y}(t)[1 - P_{0Y}(t)]}, \tag{7.22}$$

$$\sigma_{RXY}(t) = \sqrt{n_T P_{XY}(t)[1 - P_{XY}(t)]}. \tag{7.23}$$

In particular, the average molecular counts have the following stationary values:

$$\overline{N}_R = n_T \overline{P}_{00} = n_T \frac{1}{1 + \frac{n_X}{K_X} + \frac{n_Y}{K_Y} + k_C \frac{n_X}{K_X}\frac{n_Y}{K_Y}}, \tag{7.24}$$

$$\overline{N}_{RX} = n_T \overline{P}_{X0} = n_T \frac{\frac{n_X}{K_X}}{1 + \frac{n_X}{K_X} + \frac{n_Y}{K_Y} + k_C \frac{n_X}{K_X}\frac{n_Y}{K_Y}}, \tag{7.25}$$

$$\overline{N}_{RY} = n_T \overline{P}_{0Y} = n_T \frac{\frac{n_Y}{K_Y}}{1 + \frac{n_X}{K_X} + \frac{n_Y}{K_Y} + k_C \frac{n_X}{K_X}\frac{n_Y}{K_Y}}, \tag{7.26}$$

$$\overline{N}_{RXY} = n_T \overline{P}_{XY} = n_T \frac{k_C \frac{n_X}{K_X}\frac{n_Y}{K_Y}}{1 + \frac{n_X}{K_X} + \frac{n_Y}{K_Y} + k_C \frac{n_X}{K_X}\frac{n_Y}{K_Y}}. \tag{7.27}$$

A comparison with Eqs. (6.29)–(6.32) reveals that cooperativity affects the average relative abundance of receptor molecules in the (X, Y) state. With positive cooperativity ($k_C > 1$), the proportion of molecules in this state increases as compared with the no cooperativity case, while for negative cooperativity ($k_C < 1$), the proportion of molecules in the (X, Y) state decreases. Regarding the standard deviations, they also converge to stationary values that can be computed by substituting Eqs. (7.9)–(7.12) into Eqs. (7.20)–(7.23). The coefficient of variation can further be calculated from its definition $CV = \sigma/N$:

$$\overline{CV}_R = \frac{1}{\sqrt{n_T}}\sqrt{\frac{1 - \overline{P}_{00}}{\overline{P}_{00}}} = \frac{1}{\sqrt{n_T}}\left(\frac{n_X}{K_X} + \frac{n_Y}{K_Y} + k_C \frac{n_X}{K_X}\frac{n_Y}{K_Y}\right), \tag{7.28}$$

$$\overline{CV}_{RX} = \frac{1}{\sqrt{n_T}}\sqrt{\frac{1 - \overline{P}_{X0}}{\overline{P}_{X0}}} = \frac{1}{\sqrt{n_T}}\frac{1 + \frac{n_Y}{K_Y} + k_C \frac{n_X}{K_X}\frac{n_Y}{K_Y}}{\frac{n_X}{K_X}}, \tag{7.29}$$

$$\overline{CV}_{RY} = \frac{1}{\sqrt{n_T}}\sqrt{\frac{1 - \overline{P}_{0Y}}{\overline{P}_{0Y}}} = \frac{1}{\sqrt{n_T}}\frac{1 + \frac{n_X}{K_X} + k_C \frac{n_X}{K_X}\frac{n_Y}{K_Y}}{\frac{n_Y}{K_Y}}, \tag{7.30}$$

$$\overline{CV}_{RXY} = \frac{1}{\sqrt{n_T}}\sqrt{\frac{1 - \overline{P}_{XY}}{\overline{P}_{XY}}} = \frac{1}{\sqrt{n_T}}\frac{1 + \frac{n_X}{K_X} + \frac{n_Y}{K_Y}}{k_C \frac{n_X}{K_X}\frac{n_Y}{K_Y}}. \tag{7.31}$$

Observe that in all the cases, the coefficients of variation are proportional to $1/\sqrt{n_T}$. Therefore, the description in terms of the average molecular counts is accurate as long as n_T is large enough so that the coefficients of variation are negligible.

7.3 Deterministic Description

The average molecular counts $N_R(t)$, $N_{RX}(t)$, and $N_{RY}(t)$ are defined as

$$N_R(t) = \sum_{n_R, n_{RX}, n_{RY}} n_R P(n_R, n_{RX}, n_{RY}; t), \tag{7.32}$$

$$N_{RX}(t) = \sum_{n_R, n_{RX}, n_{RY}} n_{RX} P(n_R, n_{RX}, n_{RY}; t), \tag{7.33}$$

$$N_{RY}(t) = \sum_{n_R, n_{RX}, n_{RY}} n_{RY} P(n_R, n_{RX}, n_{RY}; t). \tag{7.34}$$

Moreover, from the conservation of receptor molecules, N_{RXY} is given by

$$N_{RXY}(t) = n_T - N_R(t) - N_{RX}(t) - N_{RY}(t). \tag{7.35}$$

By differentiating (7.32)–(7.34) and substituting (7.13) we can derive the differential equations that govern the dynamics of $N_R(t)$, $N_{RX}(t)$, and $N_{RY}(t)$. After carrying out all of the involved calculations we obtain

$$\frac{dN_R(t)}{dt} = -k_X^+ n_X N_R(t) + k_X^- N_{RX}(t) - k_Y^+ n_Y N_R(t) + k_Y^- N_{RY}(t), \tag{7.36}$$

$$\frac{dN_{RX}(t)}{dt} = k_X^+ n_X N_R(t) - k_X^- N_{RX}(t) - k_Y^+ n_Y N_{RX}(t) + \frac{k_Y^-}{k_C} N_{RXY}(t), \tag{7.37}$$

$$\frac{dN_{RY}(t)}{dt} = k_Y^+ n_Y N_R(t) - k_Y^- N_{RY}(t) - k_X^+ n_X N_{RY}(t) + \frac{k_X^-}{k_C} k_Y^- N_{RXY}(t), \tag{7.38}$$

with n_{RXY} as given by Eq. (7.35). Equations (7.36)–(7.38) form a complete set of differential equations. Nonetheless, from Eq. (7.35) and the differential equations above, we can also write a redundant differential equation for $N_{RXY}(t)$:

$$\frac{dN_{RXY}(t)}{dt} = k_Y^+ n_Y N_{RX}(t) - \frac{k_Y^-}{k_C} N_{RXY}(t) + k_X^+ n_X N_{RY}(t) - \frac{k_X^-}{k_C} N_{RXY}(t). \tag{7.39}$$

To analyze the above dynamical system we need to find its stationary solutions. It is not hard to prove that the system of differential equations (7.36)–(7.38) has a unique fixed point given by Eqs. (7.24)–(7.28). The stationary value for $N_{RXY}(t)$ can then be computed from Eqs. (7.24)–(7.28) and Eq. (7.35), resulting in the expression in Eq. (7.29).

It is possible to linearize the differentiate equation system (7.36)–(7.38) by making the following change of variables:

$$x(t) = N_{RX}(t) - \overline{N}_{RX}, \quad y(t) = N_{RY}(t) - \overline{N}_{RY}, \quad z(t) = N_R(t) - \overline{N}_R.$$

The differential-equation system governing the dynamics of these new variables can then be written as

$$\frac{d\mathbf{x}}{dt} = \mathbf{A}\mathbf{x}, \tag{7.40}$$

with $\mathbf{x} = (x, y, z)^T$ and

$$\mathbf{A} = \begin{bmatrix} -\left(k_X^- + k_Y^+ n_Y + \frac{k_Y^-}{k_C}\right) & -\frac{k_Y^-}{k_C} & \left(k_X^+ n_X - \frac{k_Y^-}{k_C}\right) \\ -\frac{k_X^-}{k_C} & -\left(k_Y^- + k_X^+ n_X + \frac{k_X^-}{k_C}\right) & \left(k_Y^+ n_Y - \frac{k_X^-}{k_C}\right) \\ k_X^- & k_Y^- & -(k_X^+ n_X + k_Y^+ n_Y) \end{bmatrix}. \tag{7.41}$$

From the theory of differential equations (Zill 2008), the general solution for (7.40) is

$$\mathbf{x}(t) = \sum_{i=1}^{3} C_i \mathbf{v}_i e^{\lambda_i t},$$

where $\mathbf{v}_i$ and λ_i are the eigenvectors and eigenvalues of matrix A in (7.41).

The computation of the eigenvalues and eigenvectors of matrix A does not result in short and closed algebraic expressions, as in the cooperativity-less case of Chap. 6. For that reason we do not include them here. Nevertheless, the readers are invited to verify via numerical examples that all the eigenvalues of matrix A are real and negative. This implies that the system stationary state is locally stable. If we recall that the eigenvalue with the lowest absolute value is inversely proportional to minus the system relaxation time, it is also possible to demonstrate that the existence of positive cooperativity generally increases the system relaxation time, while negative cooperativity generally decreases it.

7.4 Two Different Binding Sites, a Single Ligand

Consider the case of a receptor R with two different binding sites for the same ligand X. This system is a particular instance of the one previously studied. Consider that Y molecules are the same as X molecules, and that states $(X, 0)$ and $(0, Y)$ are equal. From the above discussion, the present system can be summarized by means of the following chemical reactions:

$$R + X \underset{k_X^- n_{RX}}{\overset{2k_X^+ n_R n_X}{\rightleftharpoons}} R_X, \tag{7.42}$$

$$R_X + X \underset{2k_X^- n_{RXX}/k_C}{\overset{k_X^+ n_{RX} n_X}{\rightleftharpoons}} R_{XX}. \tag{7.43}$$

The reaction in (7.42) results from lumping together the reactions in (7.1) and (7.2). The factor 2 in the forward reaction rate accounts for the fact that, when the receptor molecule is empty, each ligand molecule has two binding sites available. Similarly, the reaction in (7.43) results from lumping together the reactions in (7.3) and (7.4), while the factor 2 in the backward reaction rate takes into account that there two possible ways in which a fully occupied receptor can turn into a molecule with only one occupied site.

Under the supposition that the system consists of n_T receptor molecules that are bound and unbound by a constant number n_X of X ligands, its state can be determined by the counts of free and bound-by-one-ligand receptors: (n_R, n_{RX}). The number of completely occupied receptors can be calculated as:

$$n_{RXX} = n_T - n_R - n_{RX}. \tag{7.44}$$

Let $P(n_R, n_{RX}; t)$ be the probability that, at time t, the system state is (n_R, n_{RX}). Then, the master equation governing its dynamics is:

$$\begin{aligned}
\frac{dP(n_R, n_{RX}; t)}{dt} = & \; 2k_X^+ n_X (n_R + 1) P(n_R + 1, n_{RX} - 1; t) \\
& - 2k_X^+ n_X n_R P(n_R, n_{RX}; t) \\
& + k_X^- (n_{RX} + 1) P(n_R - 1, n_{RX} + 1; t) \\
& - k_X^- (n_{RX} + 1) P(n_R, n_{RX}; t) \\
& + k_X^+ n_X (n_{RX} + 1) P(n_R, n_{RX} + 1; t) \\
& - k_X^+ n_X n_{RX} P(n_R, n_{RX}; t) \\
& + 2k_X^- (n_{RXX} + 1) P(n_R, n_{RX} - 1; t) \\
& - 2k_X^- n_{RXX} P(n_R, n_{RX}; t),
\end{aligned} \tag{7.45}$$

with n_{RXX} as given by Eq. (7.44). It is straightforward to prove by substitution that the stationary solution to this master equation is

$$\overline{P}(n_R, n_{RX}) = \frac{n_T!}{n_R! n_{RX}! n_{RXX}!} \overline{P}_R^{n_R} \overline{P}_{RX}^{n_{RX}} \overline{P}_{RXX}^{n_{RXX}}, \tag{7.46}$$

in which n_{RXX} is as defined in (7.44), while

$$\overline{P}_R = \frac{1}{1 + 2\frac{n_X}{K_X} + k_C\left(\frac{n_X}{K_X}\right)^2}, \tag{7.47}$$

$$\overline{P}_{RX} = \frac{2\frac{n_X}{K_X}}{1 + 2\frac{n_X}{K_X} + k_C\left(\frac{n_X}{K_X}\right)^2}, \tag{7.48}$$

$$\overline{P}_{RXX} = \frac{k_C\left(\frac{n_X}{K_X}\right)^2}{1 + 2\frac{n_X}{K_X} + k_C\left(\frac{n_X}{K_X}\right)^2}. \tag{7.49}$$

Finally, from the properties of the multinomial distribution (Evans et al. 2000), the average numbers of molecules in every state are:

$$\overline{N}_R = n_T\frac{1}{1 + 2\frac{n_X}{K_X} + k_C\left(\frac{n_X}{K_X}\right)^2}, \tag{7.50}$$

$$\overline{N}_{RX} = n_T\frac{2\frac{n_X}{K_X}}{1 + 2\frac{n_X}{K_X} + k_C\left(\frac{n_X}{K_X}\right)^2}, \tag{7.51}$$

$$\overline{N}_{RXX} = n_T\frac{k_C\left(\frac{n_X}{K_X}\right)^2}{1 + 2\frac{n_X}{K_X} + k_C\left(\frac{n_X}{K_X}\right)^2}. \tag{7.52}$$

Let us analyze the behavior of Eqs. (7.50)–(7.52) in a few particular cases of k_C values. When there is no cooperativity, $k_C = 1$, we have

$$\overline{N}_R = n_T\frac{1}{\left(1 + \frac{n_X}{K_X}\right)^2},$$

$$\overline{N}_{RX} = n_T\frac{2\frac{n_X}{K_X}}{\left(1 + \frac{n_X}{K_X}\right)^2},$$

$$\overline{N}_{RXX} = n_T\frac{\left(\frac{n_X}{K_X}\right)^2}{\left(1 + \frac{n_X}{K_X}\right)^2}.$$

Hence, $\overline{N}_R$, $\overline{N}_{RX}$, and $\overline{N}_{RXX}$ behave as follows:

- $\overline{N}_R$ is a monotonic decreasing function of n_X: $\overline{N}_R = n_T$ when $n_X = 0$, $\overline{N}_R = n_T/2$ when $n_X = (\sqrt{2} - 1)K_X$, and $\overline{N}_R \to 0$ when $n_X \to \infty$.

- $\overline{N}_{RX}$ is a concave function of n_X: $\overline{N}_{RX} = 0$ at $n_X = 0$, the maximum value $\overline{N}_{RX} = n_T/2$ occurs when $n_X = K_X$, and $\overline{N}_{RX} \to 0$ when $n_X \to \infty$.
- $\overline{N}_{RX}$ is a monotonic increasing function of n_X: $\overline{N}_{RX} = 0$ when $n_X = 0$, $\overline{N}_R = n_T/2$ when $n_X = K_X/(\sqrt{2}-1)$, and $\overline{N}_R \to n_T$ when $n_X \to \infty$.

Other cases of interest are when k_C is either very small or very large. To analyze them it is convenient to perform the following variable change

$$x = \frac{n_X}{K_X}\sqrt{k_C}.$$

In terms of this new variable, Eqs. (7.50)–(7.52) become

$$\overline{N}_R = n_T \frac{1}{1 + \frac{2}{\sqrt{k_C}}x + x^2}, \tag{7.53}$$

$$\overline{N}_{RX} = n_T \frac{\frac{2}{\sqrt{k_C}}x}{1 + \frac{2}{\sqrt{k_C}}x + x^2}, \tag{7.54}$$

$$\overline{N}_{RXX} = n_T \frac{x^2}{1 + \frac{2}{\sqrt{k_C}}x + x^2}. \tag{7.55}$$

Thus, if $k_C \approx 0$ the quadratic term is negligible with respect to the linear one and so

$$\overline{N}_R \approx n_T \frac{1}{1 + 2\frac{n_X}{K_X}},$$

$$\overline{N}_{RX} \approx n_T \frac{2\frac{n_X}{K_X}}{1 + 2\frac{n_X}{K_X}},$$

$$\overline{N}_{RXX} \approx 0.$$

We see from the above equations that

- $\overline{N}_R$ is a monotonic decreasing function of n_X: $\overline{N}_R = n_T$ when $n_X = 0$, $\overline{N}_R = n_T/2$ when $n_X = K_X/2$, and $\overline{N}_R \to 0$ as $n_X \to \infty$.
- $\overline{N}_{RX}$ is a monotonic increasing function of n_X: $\overline{N}_{RX} = 0$ when $n_X = 0$, $\overline{N}_{RX} = n_T/2$ when $n_X = K_X/2$, and $\overline{N}_{RX} \to n_T$ as $n_X \to \infty$.
- In other words, negative cooperativity is so intense that it effectively prevents the simultaneous binding of ligands to a single receptor. The reader is invited to demonstrate that the same expressions for $\overline{N}_R$ and $\overline{N}_{RX}$ are obtained for the case in which a receptor molecule has only one binding site for X ligands, with a dissociation constant $K_X/2$. The factor $1/2$ originates from the fact that the effective association rate in a receptor with two binding sites is $2k_X^+$.

Let us now study the case in which $k_C \gg 1$. When this happens, the linear term in Eqs. (7.53)–(7.55) is negligible as compared with the quadratic one, and so Eqs. (7.50)–(7.52) can be approximated as

$$\overline{N}_R \approx n_T \frac{1}{1 + k_C \left(\frac{n_X}{K_X}\right)^2},$$

$$\overline{N}_{RX} \approx 0,$$

$$\overline{N}_{RXX} \approx n_T \frac{k_C \left(\frac{n_X}{K_X}\right)^2}{1 + k_C \left(\frac{n_X}{K_X}\right)^2}.$$

These equations lead to the following conclusions:

- $\overline{N}_R$ is a sigmoidal decreasing function of n_X: $\overline{N}_R = n_T$ at $n_X = 0$, $\overline{N}_R = n_T/2$ when $n_X = K_X/\sqrt{k_C}$, and $\overline{N}_R \to 0$ as $n_X \to \infty$.
- $\overline{N}_{RXX}$ is a sigmoidal increasing function of n_X: $\overline{N}_{RXX} = 0$ at $n_X = 0$, $\overline{N}_R = n_T/2$ when $n_X = K_X/\sqrt{k_C}$, and $\overline{N}_R \to n_T$ as $n_X \to \infty$.
- In this case, positive cooperativity is so strong that most of the time the two binding receptor sites are either occupied or unoccupied. The equations above are a particular example of the celebrated Hill equations (Santillán 2008).

7.5 Thermodynamic Analysis

To carry out the thermodynamic analysis, refer to Fig. 6.2 in particular, and to Sect. 6.5 in general. By making use of Kramer's theory (Van Kampen 1992), one can write the various reaction rates in (7.1)–(7.4) in terms of the corresponding energy levels in the energy profiles of Fig. 6.2:

$$k_X^+ = \beta e^{\mu_R^O/k_B T} e^{\mu_X^O/k_B T} e^{\mu_Y^O/k_B T} e^{-\xi_1/k_B T} \tag{7.56}$$

$$= \beta e^{\mu_{RY}^O/k_B T} e^{\mu_X^O/k_B T} e^{-\zeta_2/k_B T}, \tag{7.57}$$

$$k_X^- = \beta e^{\mu_{RX}^O/k_B T} e^{\mu_Y^O/k_B T} e^{-\xi_1/k_B T} \tag{7.58}$$

$$= \beta k_C e^{\mu_{RXY}^O/k_B T} e^{-\zeta_2/k_B T}, \tag{7.59}$$

$$k_Y^+ = \beta e^{\mu_R^O/k_B T} e^{\mu_X^O/k_B T} e^{\mu_Y^O/k_B T} e^{-\xi_2/k_B T} \tag{7.60}$$

$$= \beta e^{\mu_{RX}^O/k_B T} e^{\mu_Y^O/k_B T} e^{-\zeta_1/k_B T}, \tag{7.61}$$

$$k_Y^- = \beta e^{\mu_{RY}^O/k_B T} e^{\mu_X^O/k_B T} e^{-\xi_1/k_B T} \tag{7.62}$$

$$= \beta k_C e^{\mu_{RXY}^O/k_B T} e^{-\zeta_2/k_B T}. \tag{7.63}$$

If we solve for $\xi_1 - \zeta_2$ from Eqs. (7.56)–(7.57), do the same from Eqs. (7.58)–(7.59), equate the results and then solve for μ^O_{RXY}, we obtain:

$$\mu^O_{RXY} = \mu^O_{RX} + \mu^O_{RY} - \mu^O_R - k_B T \ln k_C. \tag{7.64}$$

The same result can be gotten by manipulating Eqs. (7.60)–(7.63).

In the present system, as in the one analyzed in the previous chapter, there is a constant number of receptor molecules (no production and no degradation), switching between the available binding states. Therefore, the system stationary state should correspond to chemical equilibrium, and so:

$$\mu_R + \mu_X + \mu_Y = \mu_{RX} + \mu_Y = \mu_{RY} + \mu_X = \mu_{RXY}. \tag{7.65}$$

Since $\mu_i = \mu_i^O + k_B T \ln \overline{N}_i$, the above equation further implies that

$$\begin{aligned}
&\overline{N}_R e^{\mu^O_R/k_B T} n_X e^{\mu^O_X/k_B T} n_Y e^{\mu^O_Y/k_B T} \\
&= \overline{N}_{RX} e^{\mu^O_{RX}/k_B T} n_Y e^{\mu^O_Y/k_B T} \\
&= \overline{N}_{RY} e^{\mu^O_{RY}/k_B T} n_X e^{\mu^O_X/k_B T} \\
&= \overline{N}_{RXY} e^{\mu^O_{RXY}/k_B T}.
\end{aligned} \tag{7.66}$$

Finally, by solving for $\overline{N}_{RX}$, $\overline{N}_{RY}$, and $\overline{N}_{RXY}$, in terms of $\overline{N}_R$ one gets:

$$\overline{N}_{RX} = \overline{N}_R \frac{n_X}{e^{(\mu^O_{RX} - \mu^O_R - \mu^O_X)/k_B T}}, \tag{7.67}$$

$$\overline{N}_{RY} = \overline{N}_R \frac{n_Y}{e^{(\mu^O_{RY} - \mu^O_R - \mu^O_Y)/k_B T}}, \tag{7.68}$$

$$\overline{N}_{RXY} = \overline{N}_R \frac{n_X n_Y}{e^{(\mu^O_{RXY} - \mu^O_R - \mu^O_X - \mu^O_Y)/k_B T}}, \tag{7.69}$$

We have from Eq. (7.65) that

$$\mu^O_{RXY} - \mu^O_R - \mu^O_X - \mu^O_Y = (\mu^O_{RX} - \mu^O_R - \mu^O_X) + (\mu^O_{RY} - \mu^O_R - \mu^O_Y) - k_B T \ln k_C. \tag{7.70}$$

That is, the energy difference between states R_{XY} and $R + X + Y$ is equal to the sum of the energy differences between states $R_X + Y$ and $R + X + Y$, plus the energy difference between states $R_Y + X$ and $R + X + Y$, plus an additional term proportional to $-\ln k_C$. In other words, cooperativity can also be interpreted from an energetic point of view: positive cooperativity increases the depth of the energy minimum corresponding to state R_{XY}, while negative cooperativity decreases it. As a matter of fact, one can define from Eq. (6.70)

$$\begin{aligned} \Delta G_X^O &= \mu_{RX}^O - \mu_R^O - \mu_X^O, \\ \Delta G_Y^O &= \mu_{RY}^O - \mu_R^O - \mu_Y^O, \\ \Delta G_{XY}^O &= \mu_{RXY}^O - \mu_R^O - \mu_X^O - \mu_Y^O. \end{aligned} \tag{7.71}$$

Then, from Eq. (7.70):

$$\Delta G_{XY}^O = \Delta G_X^O + \Delta G_Y^O + \Delta G_C, \tag{7.72}$$

with

$$\Delta G_C = -k_B T \ln k_C. \tag{7.73}$$

Let us choose the zero energy level in the energy landscape of Fig. 6.2 in such a way that $\mu_R^O + \mu_X^O + \mu_Y^0 = 0$. With this, $\Delta G_X^O, \Delta G_Y^O, \Delta G_{XY}^0 < 0$ can be interpreted as the binding energies of states $R_X + Y$, $R_Y + X$, and R_{XY}, respectively. Furthermore, Eq. (7.72) means that when the two ligands are bound to the receptor, to the total energy of the complex is the sum of the binding energies associated with each ligand, plus an extra term (ΔG_C) associated with the cooperative interaction between ligands.

By manipulating Eqs. (7.67)–(7.69), (7.70), and (7.71), and taking into account that

$$n_T = \overline{N}_R + \overline{N}_{RX} + \overline{N}_{RY} + \overline{N}_{RXY}$$

one can finally obtain

$$\overline{N}_R = n_T \frac{1}{1 + n_X e^{-\frac{\Delta G_X^O}{k_B T}} + n_Y e^{-\frac{\Delta G_Y^O}{k_B T}} + n_X n_Y e^{-\frac{\Delta G_X^O + \Delta G_Y^O + \Delta G_C}{k_B T}}}, \tag{7.74}$$

$$\overline{N}_{RX} = n_T \frac{n_X e^{-\frac{\Delta G_X^O}{k_B T}}}{1 + n_X e^{-\frac{\Delta G_X^O}{k_B T}} + n_Y e^{-\frac{\Delta G_Y^O}{k_B T}} + n_X n_Y e^{-\frac{\Delta G_X^O + \Delta G_Y^O + \Delta G_C}{k_B T}}}, \tag{7.75}$$

$$\overline{N}_{RY} = n_T \frac{n_Y e^{-\frac{\Delta G_Y^O}{k_B T}}}{1 + n_X e^{-\frac{\Delta G_X^O}{k_B T}} + n_Y e^{-\frac{\Delta G_Y^O}{k_B T}} + n_X n_Y e^{-\frac{\Delta G_X^O + \Delta G_Y^O + \Delta G_C}{k_B T}}}, \tag{7.76}$$

$$\overline{N}_{RXY} = n_T \frac{n_X n_Y e^{-\frac{\Delta G_X^O + \Delta G_Y + \Delta G_C}{k_B T}}}{1 + n_X e^{-\frac{\Delta G_X^O}{k_B T}} + n_Y e^{-\frac{\Delta G_Y^O}{k_B T}} + n_X n_Y e^{-\frac{\Delta G_X^O + \Delta G_Y^O + \Delta G_C}{k_B T}}}. \tag{7.77}$$

We can see in the above equations that when $k_C > 1$ (positive cooperativity), $\Delta G_C < 0$ and so the binding energy of the complex R_{XY} is more negative than the sum of individual binding energies, hereby increasing its stability and the proportion of $\overline{N}_{RXY}$ complexes. On the contrary, $k_C < 1$ (negative cooperativity) implies that the binding energy of R_{XY} is less negative than the sum of the individual binding energies, and predicts a decrement in the proportion of $\overline{N}_{RXY}$ complexes.

Let us rewrite Eq. (6.78):

$$e^{-\Delta G_X^O/k_B T} = k_X^+/k_X^- = 1/K_X, \quad e^{-\Delta G_Y^O/k_B T} = k_Y^+/k_Y^- = 1/K_y.$$

This last expression, together with Eq. (7.73), allows to recast Eqs. (7.74)–(7.77) as

$$\overline{N}_R = n_T \frac{1}{1 + \frac{n_X}{K_X} + \frac{n_Y}{K_Y} + k_C \frac{n_X}{K_X}\frac{n_Y}{K_Y}},$$

$$\overline{N}_{RX} = n_T \frac{\frac{n_X}{K_X}}{1 + \frac{n_X}{K_X} + \frac{n_Y}{K_Y} + k_C \frac{n_X}{K_X}\frac{n_Y}{K_Y}},$$

$$\overline{N}_{RY} = n_T \frac{\frac{n_Y}{K_Y}}{1 + \frac{n_X}{K_X} + \frac{n_Y}{K_Y} + k_C \frac{n_X}{K_X}\frac{n_Y}{K_Y}},$$

$$\overline{N}_{RXY} = n_T \frac{k_C \frac{n_X}{K_X}\frac{n_Y}{K_Y}}{1 + \frac{n_X}{K_X} + \frac{n_Y}{K_Y} + k_C \frac{n_X}{K_X}\frac{n_Y}{K_Y}},$$

which are nothing but Eqs. (7.25)–(7.27). This proves the equivalence between the dynamic and thermodynamic approaches, and allows to build connections between thermodynamic and chemical kinetics concepts like free energies, reaction rates, and dissociation constants.

The results in Sect. 7.4 can be restated as follows from a thermodynamic perspective. Assume that the receptor molecules have two binding sites for the same ligand, and that the ligands interact cooperatively as described above. Let us consider the following extreme cases for the cooperativity constant: $k_C \approx 0$ and $k_C \gg 1$. Notice from Eq. (7.73) that $k_C \approx 0$ implies that $\Delta G_C \gg 0$. Therefore, the binding energy of the state R_{XX} (which in this case is dominated by ΔG_C) will be much larger than that of the other states, making it extremely unstable and unlikely. Contrarily, $k_C \gg 1$ implies that $\Delta G_C \ll 0$. This means that state R_{XX} has an energy which is much lower than that of state R_X, and consequently the number of molecules in state R_X is very small as compared to that in R_{XX}.

7.6 Summary

This chapter generalizes the results in Chap. 6 by considering cooperativity. In here make use of all the previously introduced concepts and techniques (both mathematical and physical). In my opinion, the results here presented speak by themselves. I only wish to emphasize that they are a very nice example of how combining the thermodynamics and dynamics approaches can greatly help to better understand a subtle concept, like cooperativity.

Chapter 8
Gene Expression and Regulation

Abstract In this chapter we make use of all the material studied so far to construct a few (simple but informative) models for gene expression. As before, we are interested in obtaining useful information regarding both the dynamical and thermodynamical aspects of this phenomenon.

8.1 A Model for Constitutive Gene Transcription

Transcription initiates when a polymerase molecule specifically binds to a small DNA region preceding the gene (the promoter) forming the so-called closed complex. Then, the polymerase opens the DNA double helix forming the open complex, and starts moving in the 5'-3' direction, assembling the RNA molecule complementary to the gene-coding DNA segment. The processes involving formation of the open complex, elongation of the nascent RNA molecule, and termination of transcription are usually lumped as a single irreversible chemical event from the standpoint of mathematical modeling (Shahrezaei and Swain 2008; Zeron and Santillán 2010). With this in mind, the processes underlying transcription can be represented in terms of chemical reactions as follows:

$$D + P \underset{k_P^- n_{DP}}{\overset{k_P^+ n_D n_P}{\rightleftharpoons}} D_P, \tag{8.1}$$

$$D_P + S \overset{k_M n_{DP} n_S}{\rightharpoonup} D + P + M. \tag{8.2}$$

In the above reactions D represents an empty promoter, D_P denotes a promoter bound by a polymerase (open complex), S corresponds to the substrates out of which the RNA molecule is made, and M is the resulting RNA molecule.

M. Santillán, *Chemical Kinetics, Stochastic Processes, and Irreversible Thermodynamics*, Lecture Notes on Mathematical Modelling in the Life Sciences,
DOI 10.1007/978-3-319-06689-9_8,

Furthermore, n_D, n_P, n_{DP}, and n_M stand for the free promoters, polymerase, polymerase-promoter closed complex, and RNA molecule counts.

The closed complex is quite unstable as compared with the open one (McClure 1985): a polymerase binds and unbinds several times to the promoter before forming the open complex. This means that the rates corresponding to the chemical reaction in (8.1) are much larger than those corresponding to the reaction in (8.2): $k_M \ll k_P^+, k_P^-$.

If we further include the reaction accounting for RNA degradation (which is also generally regarded as irreversible):

$$M \overset{\gamma_M n_M}{\rightharpoonup} Y, \tag{8.3}$$

and assume that the RNA degradation rate is such that $\gamma_M \ll k_P^+, k_P^-$, then the chemical-reaction system (8.1)–(8.3) happens to be identical to the one studied in Chap. 5—See Eqs. (5.7)–(5.9). In this case, the promoter takes the place of the enzyme, polymerase molecules P take the place of the global reaction substrate, and M is the product of the enzymatic reaction. Henceforth, all the obtained conclusions regarding the dynamic and thermodynamic behavior of the system in (5.7)–(5.9) can be mapped as follows:

- The assumed inequalities $\gamma_M, k_M \ll k_P^+, k_P^-$ allow a quasi-stationary approximation that consists in splitting the whole system into a fast subsystem (the promoter flipping back and forth between the free and the closed complex states), nested into a slow subsystem (RNA production and degradation). In this approximation, the fast subsystem equilibrates instantaneously with the slow one, which dictates the temporal evolution of the whole system.
- In the stationary state, the probability that the promoter is bound by a polymerase is

$$\frac{n_P}{n_P + K_P}, \tag{8.4}$$

  with $K_P = k_P^-/k_P^+$.
- Regarding the *RNA* molecule count, the corresponding stationary probability distribution is a Poisson distribution with mean value equal to

$$\overline{N}_M = \frac{k_M}{\gamma_M} \frac{n_P}{n_P + K_P}. \tag{8.5}$$

  Recall that the variance in a Poisson distribution equals the mean value.
- The relaxation time of the fast subsystem is proportional to $(k_P^+ + k_P^-)^{-1}$, while the relaxation time of the slow subsystem is proportional to γ_M^{-1}. Thus, the inequality $\gamma_M \ll k_P^+, k_P^-$ guaranties a clear separation of scales in the relaxation times.

- The fast subsystem stationary state complies with chemical equilibrium, which is characterized by states D and D_P having equal chemical potentials. Concomitantly, there is no net flux of chemical energy associated with the fast subsystem.
- The stationary state of the slow subsystem does not comply with chemical equilibrium. Consequently the chemical potentials of states $D_P + S$ and $D + P + M$ obey the following inequality

$$\mu_{DP} + \mu_S > \mu_D + \mu_P + \mu_M. \tag{8.6}$$

However, since $\mu_{DP} = \mu_D + \mu_P$, it follows from the above inequality that

$$\mu_S > \mu_M. \tag{8.7}$$

That is, the substrates out of which RNA molecules are made have a higher chemical potential than RNA molecules themselves. Thanks to this difference there exists a nonzero rate of conversion of S molecules into RNAs given by

$$k_M^{\text{eff}} = k_M n_S \frac{n_P}{n_P + K_P}. \tag{8.8}$$

There exists also a chemical difference between state M and state Y (which represents the products of RNA degradation): $\mu_M > \mu_Y$. Furthermore, the RNA degradation rate equals the rate of production of this molecule in the stationary state. From all these considerations, the rate of heat dissipation in this system is

$$\Phi = k_M^{\text{eff}}(\mu_S - \mu_M) + k_M^{\text{eff}}(\mu_M - \mu_Y) = k_M^{\text{eff}}(\mu_S - \mu_Y). \tag{8.9}$$

In accordance to the first law of thermodynamics, energy has to be supplied to the system at the same rate to keep it constant. This occurs through the constant addition of S molecules and the constant removal of Y molecules.
- The differential equation governing the dynamics of the mean RNA count is

$$\frac{dN_M(t)}{dt} = k_M \frac{n_P}{n_P + K_P} - \gamma_M N_M(t). \tag{8.10}$$

8.2 A Fast-Regulation Model for Transcription

Often times, gene expression is regulated at the transcriptional level. The first discovered transcriptional regulatory mechanism was repression, and although many more regulatory mechanisms have been discovered since then, repression remains one of the most common ones. In this mechanism, a molecule known as the repressor binds the promoter in a specific site and prevents transcription initiation, even if a polymerase is bound to its corresponding site. The following reactions account for this regulatory process:

$$D + P \underset{k_P^- n_{DP}}{\overset{k_P^+ n_D n_P}{\rightleftharpoons}} D_P, \tag{8.11}$$

$$D + R \underset{k_R^- n_{DR}}{\overset{k_R^+ n_D n_R}{\rightleftharpoons}} D_R, \tag{8.12}$$

$$D_R + P \underset{k_P^- n_{DPR}}{\overset{k_P^+ n_{DR} n_P}{\rightleftharpoons}} D_{PR}, \tag{8.13}$$

$$D_P + R \underset{k_R^- n_{DPR}}{\overset{k_R^+ n_{DP} n_R}{\rightleftharpoons}} D_{PR}, \tag{8.14}$$

$$D_P + S \overset{k_M n_{DP} n_S}{\rightharpoonup} D + P + M, \tag{8.15}$$

$$M \overset{\gamma_M n_M}{\rightharpoonup} Y. \tag{8.16}$$

The newly introduced variables are: R, which represents a repressor molecule; D_R, that stands for the promoter–repressor complex; D_{PR}, which corresponds to the promoter–polymerase–repressor complex; n_R, n_{DR}, n_{DPR}, representing the corresponding molecular counts; and k_R^+ and k_R^-, the association and dissociation rate constants for the binding and unbinding of the repressor to its corresponding site on the promoter.

Under the assumption that the binding and unbinding of repressor and polymerase from the promoter are much faster processes than the synthesis and degradation of RNA, one can make a quasi-stationary approximation similar to the one we made in the previous section. According to this approximation we can split the system into fast and slow subsystems. The fast subsystem comprises the binding and unbinding of repressor and polymerase molecules to the promoter, while the slow subsystem accounts for the synthesis and degradation of RNA. Moreover, from the slow subsystem perspective, the fast subsystem reaches stationarity instantaneously.

The fast subsystem is identical to that studied in Chap. 6. Hence, the stationary state of the fast subsystem complies with chemical equilibrium and so the states $D + R + P$, $D_R + P$, $D_P + R$, and D_{RP} have all the same chemical potential. Furthermore, the stationary probability of finding the promoter bound by a polymerase and free from any repressor is

$$\frac{n_P/K_P}{(1 + n_P/K_P)(1 + n_R/K_R)} = \frac{k_R}{n_R + K_R} \frac{n_P}{n_P + K_P}, \tag{8.17}$$

with $K_R = k_R^-/k_R^+$. Since the promoter state D_P is the only one out of which transcription can start, Eq. (8.17) is directly comparable with Eq. (8.4). We see that the presence of the repressor modulates the probability that the promoter is in the transcriptionally active state. The more repressor molecules are present, the smaller this probability. The dissociation constant K_R determines the number of repressor molecules at which the probability that the promoter is in the transcriptionally active state decrease 50 %.

Regarding RNA, its dynamics are quite similar to those of the constitutive transcription. RNA molecules are synthesized via a Poisson process with rate

$$k_M^{\text{eff}} = k_M \frac{K_R}{n_R + K_R} \frac{n_P}{n_P + K_P}, \tag{8.18}$$

and are degraded linearly. In consequence, the stationary probability distribution for the RNA molecular count is a Poisson distribution with mean

$$\overline{N}_M = \frac{k_M}{\gamma_M} \frac{K_R}{n_R + K_R} \frac{n_P}{n_P + K_P}. \tag{8.19}$$

The synthesis and degradation of RNA molecules is a process which, even in the stationary state, is out of equilibrium. The reason is the unbalance of chemical potentials between the substrates out of which RNA is synthesized and the pool of RNA molecules, and between the RNA pool and the products of RNA degradation. These unbalances further implies the existence of a heat dissipation rate given by

$$\Phi = k_M^{\text{eff}}(\mu_S - \mu_Y), \tag{8.20}$$

with k_M^{eff} as given by Eq. (8.18).

Finally, it is not hard to prove that the differential equation governing the dynamics of the average count of RNA molecules, $N(t)$ is:

$$\frac{dN_M(t)}{dt} = k_M \frac{k_R}{n_R + K_R} \frac{n_P}{n_P + K_P} - \gamma_M N_M(t). \tag{8.21}$$

A comparison of the above results with those of the previous section reveals that the sole effect of adding repressors is decreasing the effective rate for RNA synthesis. Other than that, the models for constitutive transcription and for regulated transcription with rapid repressor–promoter interaction behave in the same way.

8.3 A Slow-Regulation Model for Transcription

In the above section we studied a model that could nicely explain how the expression of a gene can be controlled by modifying the number of regulatory molecules: repressors. The problem with this model is that it relies upon an assumption that

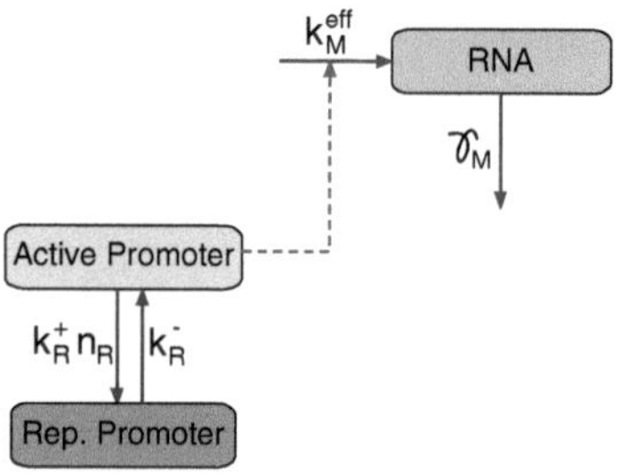

Fig. 8.1 Schematic representation of a reduced model for gene expression with slow interaction between repressor molecules and DNA promoter

has been proven invalid by recent experimental results (Golding et al. 2005). As a matter of fact repressor binding and unbinding is the slowest process in the reaction scheme in (8.11)–(8.16). Thus, there exist three different time scales. The fastest one corresponds to the closed complex formation and dissociation processes, in a slower scale we have the RNA synthesis and degradation processes, and finally, the slowest time scale is that of repressor binding and unbinding.

Being promoter binding and unbinding by a polymerase the fastest processes of gene expression, we can make a quasi-stationary approximation similar the ones we have done in the previous sections. As a result we get a reduced system schematically represented in Fig. 8.1. The chemical reactions governing the dynamics of this system are then as follows:

$$D + R \underset{k_R^- n_{DR}}{\overset{k_R^+ n_R n_D}{\rightleftharpoons}} D_R \tag{8.22}$$

$$S \overset{k_M^{\text{eff}} n_D}{\rightharpoonup} M \tag{8.23}$$

$$M \overset{\gamma_M n_M}{\rightharpoonup} Y, \tag{8.24}$$

with $k_M^{\text{eff}} = k_M n_P/(n_P + k_P)$. The master equation corresponding to these reactions, with the assumption that only one promoter exists ($n_D + n_{DR} = 1$), is

$$\begin{aligned}\frac{dP(n_D, n_M; t)}{dt} &= k_R^- n_D P(1 - n_D, n_M; t) - k_R^-(1 - n_D) P(n_D, n_M; t)\\ &+ k_R^+(1 - n_D) n_R P(1 - n_D, n_M; t) - k_R^+ n_D n_R P(n_D, n_M; t)\\ &+ k_M^{\text{eff}} n_D P(n_D, n_M - 1; t) - k_M^{\text{eff}} n_D P(n_D, n_M; t)\\ &+ \gamma_M n_M P(n_D, n_M + 1; t) - \gamma_M n_M P(n_D, n_M; t)\end{aligned} \tag{8.25}$$

The characteristic times of the RNA synthesis and degradation processes range from a few seconds to a few minutes, while the characteristic times for promoter activation and deactivation are of the order of tens of minutes. In other words, there is a separation of time scales of one order of magnitude (Zeron and Santillán 2010). Typically, this is not enough to support a quasi-stationary approximation. However, for the sake of simplicity we shall make this approximation in the understanding that the obtained conclusions should be taken cautiously. Observe in (8.25) that the terms corresponding to promoter activation and repression (first row on the right-hand side) do not involve the number of RNA molecules (n_M). Conversely, the terms corresponding to the synthesis of RNA molecules (second row on the right-hand side) do depend on the number of active promoters. Therefore, in this case the slow processes take place independently of the fast ones. The slow processes (promoter activation and repression) are equivalent to the system studied in Chap. 3. This allows us to derive the following conclusions:

- The promoter flips back and forth between the repressed and the active states with the following repression and activation rates: $k_R^+ n_R$ and k_R^-.
- The probability distribution for the times during which the promoter remains repressed is an exponential distribution with mean $(k_R^-)^{-1}$.
- The probability that the promoter is repressed at any given time is $n_R/(n_R + K_R)$.
- The probability distribution for the times during which the promoter remains active is an exponential distribution with mean $(k_R^+ n_R)^{-1}$. We see that, in this case, the effect of increasing the number of repressors is to decrease the promoter average active time.
- The probability that the promoter is active at any given time is $K_R/(n_R + K_R)$. In agreement with the above assertion, increasing the number of repressors decreases the probability that the promoter is active.

Regarding RNA dynamics, given that it has the fastest time scale, it follows the dynamics of promoter activation and repression instantaneously. Under the assumption that a quasi-stationary approximation is valid, one can assert that the population of *RNA* molecules almost immediately extinguishes when the promoter becomes repressed, and that it rapidly evolves to a stationary population with a Poisson distribution when the promoter is active. More specifically, the RNA population jumps from no molecules when the promoter is repressed, to a fluctuating population obeying a Poisson distribution with mean $\lambda = k_M^{\text{eff}}/\gamma_M$ when the promoter is active. Given that the probability that the promoter is active at any given time is $K_R/(n_R + K_R)$, the stationary probability of finding n_M RNA molecules at time t is:

$$P(n_M) = \frac{n_R}{n_R + K_R}\delta_{n_M 0} + \frac{K_R}{n_R + K_R}\frac{\lambda^{n_M} e^{-\lambda}}{n_M!}, \tag{8.26}$$

with $\delta_{n_M 0}$ Kroneker's delta. This result can be derived from Eq. (8.25) by taking into account that $P(n_M;t) = \sum_{n_D=0}^{1} P(n_D, n_M;t)$, imposing the quasi-stationary approximation, and assuming that the whole system is in a stationary state. When compared with the Poisson distribution obtained for the system studied in the last section, the distribution in (8.26) is wider and in some instances can be bimodal. Interestingly, the mean value for this probability distribution is

$$\overline{N}_M = \sum_{n=0}^{\infty} n_M P(n_M) = \frac{k_M}{\gamma_M} \frac{K_R}{n_R + K_R} \frac{n_P}{n_P + K_P},$$

which is identical to the expression in (8.19). This means that, although the stochastic dynamics are quite different, the mean stationary average number of RNA molecules obtained from the models with fast and slow promoter repression-activation dynamics are the same.

By definition, the average number of active promoters and of RNA molecules are

$$N_D(t) = \sum_{n_D, n_M} n_D P(n_D, n_M; t)$$

$$N_M(t) = \sum_{n_D, n_M} n_M P(n_D, n_M; t)$$

By differentiating these expressions and substituting the master equation in (8.25) we obtain the following differential equations

$$\frac{dN_D(t)}{dt} = k_R^-(1 - N_D(t)) - k_R^+ n_R N_D(t), \tag{8.27}$$

$$\frac{dN_M(t)}{dt} = k_M \frac{n_P}{n_P + K_P} N_D(t) - \gamma_M N_M(t). \tag{8.28}$$

When we compare with Eq. (8.21) we see that the average counts of the system with slow promoter repression-activation kinetics follow quite a different dynamics, contrasted with the system with fast repressor kinetics. However, after a little algebra we get the following stationary values

$$\overline{N}_D = \frac{K_R}{n_R + K_R}, \tag{8.29}$$

$$\overline{N}_M = \frac{k_M}{\gamma_M} \frac{K_R}{n_R + K_R} \frac{n_P}{n_P + K_P}, \tag{8.30}$$

which are completely compatible with the results of the fast repressor system—see Eq. (8.19).

What about the thermodynamic considerations? As we have seen, processes that involve the transition of a fixed number of molecules between different states are compatible with thermodynamic equilibrium in the steady state, because all the states have the same chemical potential. The contrary happens with processes in which molecules are constantly produced and degraded. The synthesis and degradation of molecules implies the conversion of high energy substrates into low energy products, as well as the dissipation as heat of the energy difference. Furthermore, in order to maintain the system stationary state, new substrate molecules have to be added to the system, while product molecules ought to be removed. This matter flux also conveys an influx of energy into the system that compensates heat dissipation. We can conclude from the above considerations that the promoter flipping between the active and inactive states does not entail any energy flux. However, there is an energy flux (heat dissipation rate) associated with RNA synthesis and degradation given by

$$\Phi = v(\mu_S - \mu_Y),$$

with v the RNA synthesis rate, which is zero when the promoter in repressed, and equals $k_M n_P/(n_P + K_P)$ when the promoter is active. That is, the cell (assuming it has a single promoter) fluctuates between no-energy-flux and high energy dissipation states. Since the promoter remains in the active state a fraction of the time equal to $K_R/(n_R + K_R)$, the average energy dissipation rate in the long run is

$$\Phi = k_M \frac{K_R}{n_R + K_R} \frac{n_P}{n_P + K_P} (\mu_S - \mu_Y).$$

The same as in the fast repressor dynamics model, see Eq. (8.20).

8.4 Gene Expression (Transcription and Translation): Stochastic Description

Gene expression in many cases involves not only transcription but also translation, which means the synthesis of protein molecules by using RNA molecules as blueprints. Translation is carried out by ribosomes, which bind the RNA at specific site, and then start traveling along this molecule, reading the genetic information in it, and assembling a protein molecule accordingly. In many senses, polymerases and ribosomes play equivalent rules. Both bind and unbind several times from their binding sites and, once in a while, they initiate their respective process (transcription or translation). Under the supposition that the binding and unbinding of polymerases and ribosomes are much faster as compared with transcription and translation initiation, gene expression in a constitutive promoter, which is schematically represented in Fig. 8.2, can be summarized by means of the following chemical reactions:

$$S_M \overset{k_M}{\rightharpoonup} M, \tag{8.31}$$

$$M \overset{\gamma_M n_M}{\rightharpoonup} Y_M, \tag{8.32}$$

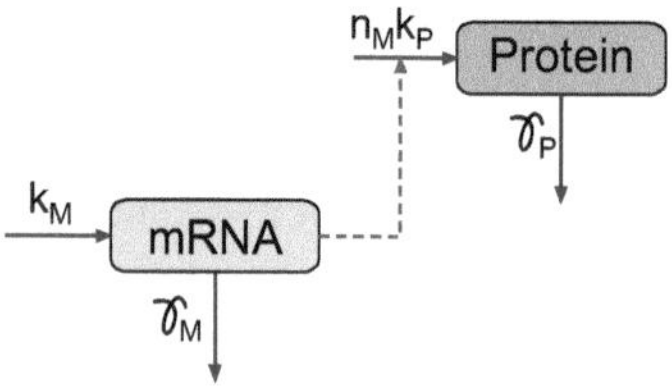

Fig. 8.2 Schematic representation of the transcription and translation processes in a constitutive gene

$$S_P \overset{k_P}{\rightharpoonup} P, \tag{8.33}$$

$$P \overset{\gamma_P n_P}{\rightharpoonup} Y_P. \tag{8.34}$$

In the reactions above M and P respectively denote messenger RNA and protein, while n_M and n_P are the corresponding molecular counts; S_M and S_P are the substrates out of which RNA and protein molecules are made of, Y_M and Y_P correspond to the products into which RNA and proteins are degraded, k_M and k_P are the RNA and protein effective synthesis rates (accounting for the abundance of substrate molecules as well as of polymerases or ribosomes), and finally γ_M and γ_P are the RNA and protein degradation rates.

The master equation corresponding to the reactions in (8.31)–(8.34) is

$$\begin{aligned}\frac{dP(n_M,n_P;t)}{dt} &= k_M P(n_M-1,n_P;t) - k_M P(n_M,n_P;t)\\ &+ \gamma_M(n_M+1)P(n_M+1,n_P;t) - \gamma_M n_M P(n_M,n_P;t)\\ &+ k_P n_M P(n_M,n_P-1;t) - k_P n_M P(n_M,n_P;t)\\ &+ \gamma_P(n_P+1)P(n_M,n_P+1;t) - \gamma_P n_P P(n_M,n_P;t),\end{aligned} \tag{8.35}$$

with $P(n_M,n_P;t)$ the probability of having n_M RNA and n_P protein molecules at time t.

If we assume that the RNA dynamics (production and degradation) are much faster than the protein dynamics, then we can make a quasi-stationary approximation as in Chap. 5. After doing it, we obtain two separated master equations, one for the RNA and the other for the protein dynamics:

$$\begin{aligned}\frac{dP(n_M;t)}{dt} &= k_M P(n_M-1;t) - k_M P(n_M;t)\\ &+ \gamma_M(n_M+1)P(n_M+1;t) - \gamma_M n_M P(n_M;t),\end{aligned} \tag{8.36}$$

$$\begin{aligned}\frac{dP(n_P;t)}{dt} &= k_P \overline{N}_M P(n_P-1;t) - k_P \overline{N}_M P(n_M;t)\\ &+ \gamma_P(n_P+1)P(n_P+1;t) - \gamma_P n_P P(n_P;t).\end{aligned} \tag{8.37}$$

In the above equations $P(n_M;t) = \sum_{n_P} P(n_M, n_P; t)$ is the probability of having n_M RNA molecules at time t, $P(n_P;t) = \sum_{n_M} P(n_M, n_P; t)$ is the probability of having n_P protein molecules at time t, and $\overline{N}_M$ is the mean number of RNA molecules predicted by the stationary solution of Eq. (8.36), which is the following Poisson distribution

$$\overline{P}(n_M) = \frac{\overline{N}_M^{n_M} e^{-\overline{N}_M}}{n_M!}, \tag{8.38}$$

with

$$\overline{N}_M = \frac{k_M}{\gamma_M}. \tag{8.39}$$

From this, the stationary solution of Eq. (8.37) is

$$\overline{P}(n_P) = \frac{\overline{N}_P^{n_P} e^{-\overline{N}_P}}{n_P!}, \tag{8.40}$$

in which the stationary average protein count, $\overline{N}_P$ is given by

$$\overline{N}_P = \frac{k_P \overline{N}_M}{\gamma_P} = \frac{k_M k_P}{\gamma_M \gamma_P}. \tag{8.41}$$

Despite its elegance and simplicity, the above analyzed approximation is incorrect because it relays on a wrong assumption. Although the protein decaying process is indeed much slower than RNA synthesis and decay, protein synthesis is as fast, if not faster, than the RNA-related processes. In what follows we shall construct a better approximation, but in order to do so it is useful to analyze Eqs. (8.36) and (8.37) in a more detailed way. In the model in (8.31)–(8.34) the RNA production and degradation processes are completely independent of the protein count, thus we expect the master equation for $P(n_M;t)$ to remains the same in the improved model as in Eq. (8.36). Regarding Eq. (8.37), in order to interpret it, it is more convenient to write it as

$$\begin{aligned}\frac{dP(n_P;t)}{dt} &= k_M \frac{k_P}{\gamma_M} P(n_P - 1; t) - k_M \frac{k_P}{\gamma_M} P(n_M; t) \\ &\quad + \gamma_P (n_P + 1) P(n_P + 1; t) - \gamma_P n_P P(n_P; t).\end{aligned}$$

Let us focus on the factor $k_M(k_P/\gamma_M)$, appearing in the first row. This factor, which accounts for the rate of protein synthesis, can be split as the rate of RNA production, k_M, times the probability that a single RNA is translated once during its life time, k_P/γ_M. To understand this last assertion, notice that translation and degradation are carried out by two different molecules that compete for the RNA:

the polymerase and the degradosome. Thus, the probability that a polymerase binds the RNA molecule before a degradosome and starts translation is

$$\frac{k_P}{k_P + \gamma_M}.$$

Furthermore, the probability that the RNA is translated r times before being degraded is

$$P(r) = \left(\frac{k_P}{k_P + \gamma_M}\right)^r \left(1 - \frac{k_P}{k_P + \gamma_M}\right). \tag{8.42}$$

Finally, under the supposition that $k_P \ll \gamma_M$, we have that $P(1) \approx k_P/\gamma_M$, and $P(r) \approx 0$ for all $r > 1$. In other words, assuming $k_P \ll \gamma_M$ is equivalent as supposing that the RNA degradation is so fast that this molecule is translated at most once during its lifetime.

What happens if the condition $k_P \ll \gamma_M$ is not fulfilled? With the aid of Eq. (8.42) the master equation for $P(n_P;t)$ can be generalized as follows

$$\begin{aligned}\frac{dP(n_P;t)}{dt} &= k_M \sum_{r=1}^{n_P} P(r)P(n_P - r;t) - k_M \sum_{r=1}^{\infty} P(r)P(n_P;t)\\ &+ \gamma_P(n_P + 1)P(n_P + 1;t) - \gamma_P n_P P(n_P;t).\end{aligned} \tag{8.43}$$

The first sum on the right-hand side of the previous equation goes from $r = 1$ to $r = n_P$ because the corresponding term accounts for all the translation events that take the system into the state with n_P proteins. Hence, since no negative protein count exists, the maximum possible number of translations is n_P. The master equation in Eq. (8.43) has been studied elsewhere (Shahrezaei and Swain 2008) and its stationary solution has been proven to be

$$\overline{P}(n_P) = \frac{\Gamma(\alpha + n_P)}{\Gamma(n_P + 1)\Gamma(\alpha)} \left(\frac{k_P}{k_P + \gamma_M}\right)^{n_P} \left(1 - \frac{k_P}{k_P + \gamma_M}\right)^{\alpha}, \tag{8.44}$$

in which $\alpha = k_M/\gamma_P$ and $\Gamma(x)$ is the gamma function. The distribution in (8.44) is a negative binomial distribution. It is immediate to obtain from the properties of this distribution (Evans et al. 2000) the following expression for the average n_P value:

$$\overline{N}_P = \sum_{\infty}^{n_P=0} n_P \overline{P}(n_P) = \frac{k_M k_P}{\gamma_M \gamma_P}, \tag{8.45}$$

which is the same as in Eq. (8.41). That is, assuming a fast translation initiation does note change the expected average protein count. However, when we compare the distribution probabilities in Eqs. (8.41) and (8.45), it results that the later presents a larger dispersion around the mean value.

Another interested example that we shall not address here due to the complexity of the mathematics involved is that in which the promoter is regulated. However, the interested readers can study it in the specialized literature.

8.5 Gene Expression: Deterministic Description

The average number of RNA and protein molecules can be computed from the respective probability distributions as follows:

$$N_M(t) = \sum_{n_M=0}^{\infty} n_M P(n_M;t), \quad N_P(t) = \sum_{n_P=0}^{\infty} n_P P(n_P;t).$$

From these definitions we can derive the differential equations governing the dynamics of $N_M(t)$ and $N_P(t)$ as

$$\frac{dN_M(t)}{dt} = \sum_{n_M=0}^{\infty} n_M \frac{dP(n_M;t)}{dt}, \quad \frac{dN_P(t)}{dt} = \sum_{n_P=0}^{\infty} n_P \frac{dP(n_P;t)}{dt}. \tag{8.46}$$

Given that the master equation for $P(n_M;t)$ is the same for the two models studied in the previous section, substitution of Eq. (8.36) into the expression for $dN_M(t)/dt$ gives

$$\frac{dN_M(t)}{dt} = k_M - \gamma_M N_M(t), \tag{8.47}$$

which has the stationary solution

$$\overline{N}_M = \frac{k_M}{\gamma_M},$$

in agreement with (8.39). Regarding the dynamics of $N_P(t)$, we have two models and so two different expressions for $dP(n_P;t)/dt$: Eqs. (8.37) and (8.43). Substitution of Eq. (8.37) into the definition of $N_P(t)$ leads to:

$$\frac{dN_P(t)}{dt} = \frac{k_M k_P}{\gamma_M} - \gamma_P N_P(t). \tag{8.48}$$

However, we know that Eq. (8.37) is not the best approximation because it relies on an assumption not supported by experimental evidence. To avoid this problem let us consider Eq. (8.43) instead. However, before doing so it is convenient to rewrite this equation as

$$\frac{dP(n_P;t)}{dt} = k_M \left(1 - \frac{k_P}{k_P + \gamma_M}\right) \sum_{r=0}^{n_P} \left(\frac{k_P}{k_P + \gamma_M}\right)^r P(n_P - r;t) - k_M P(n_P;t)$$
$$+ \gamma_P (n_P + 1) P(n_P + 1;t) - \gamma_P n_P P(n_P;t). \tag{8.49}$$

In the derivation of this last equation we have added and subtracted the term

$$\left(1 - \frac{k_P}{k_P + \gamma_M}\right) P(n_P;t)$$

and made use of the equality

$$\sum_{n=0}^{\infty} a^n = \frac{1}{1+a}, \quad \text{if} \quad |a| < 1. \tag{8.50}$$

By substituting (8.49) into (8.46) we get

$$\frac{dN_P(t)}{dt} = k_M \left(1 - \frac{k_P}{k_P + \gamma_M}\right) \sum_{r=0}^{n_P} \left(\frac{k_P}{k_P + \gamma_M}\right)^r \sum_{n_P=0}^{\infty} n_P P(n_P - r;t) - k_M N_P(t)$$
$$+ \gamma_P \sum_{n_P=0}^{\infty} n_P (n_P + 1) P(n_P + 1;t) - \gamma_P N_P(t)^2.$$

If we make the index substitution $n = n_P - 1$ in the summation over n_P appearing in the first row of the equation above, and the substitution $n = n_P + 1$ in the summation over n_P appearing in the second row, we get:

$$\frac{dN_P(t)}{dt} = k_M \left(1 - \frac{k_P}{k_P + \gamma_M}\right) \sum_{r=0}^{n_P} \left(\frac{k_P}{k_P + \gamma_M}\right)^r (N_P(t) + r) - k_M N_P(t)$$
$$+ \gamma_P N_P(t)(N_P(t) - 1) - \gamma_P N_P(t)^2.$$

Finally, by simplifying, making use of the equality in (8.50), and taking into account that

$$\sum_{n=0}^{\infty} n a^n = \frac{a}{(a-1)^2}, \quad \text{if} \quad |a| < 1,$$

we obtain

$$\frac{dN_P(t)}{dt} = \frac{k_M k_P}{\gamma_M} - \gamma_P N_P(t),$$

which is the same result as Eq. (8.48). In other words, the deterministic description is the same no matter whether one makes the incorrect assumption that the protein synthesis rate is much slower that the RNA production and degradation rates or not.

8.6 Gene Expression: Thermodynamic Interpretation

As we have previously seen, the only processes that contribute to net energy flux and heat dissipation are those in which molecules are produced and degraded, and the net chemical potential of the substrates is higher than that of the end products. On the other hand, the energy fluxes depend on the average rate of molecule synthesis or degradation (in the stationary state, the production and degradation rates are equal). Hence, given that the deterministic descriptions for the two studied models are identical, the computed heat dissipation rates are the same in both cases. The net heat rate consists of a part associated with RNA synthesis and degradation:

$$\Phi_M = k_M(\mu_{SM} - \mu_{YM}),$$

plus another part associated with protein synthesis and degradation:

$$\Phi_P = k_M \frac{k_P}{\gamma_M}(\mu_{SP} - \mu_{YP}).$$

In the equations above μ_{SM} and μ_{SP} respectively represent the chemical potentials of the substrates out of which RNA and proteins are made. On the other hand, μ_{YM} and μ_{YP} are the chemical potentials of the final products into which RNA and proteins are degraded.

Although we did not study the dynamics of gene regulation when the promoter is regulated, we do can compute the heat dissipation rate. The reason is that, as we have seen, promoter regulation is a process that complies with chemical equilibrium, and so does not contribute to the heat dissipation rate. Therefore, we only need to take into account the effect of repressors on transcription to obtain:

$$\Phi_M = k_M \frac{K_R}{n_R + K_R}(\mu_{SM} - \mu_{YM}),$$

and

$$\Phi_P = k_M \frac{K_R}{n_R + K_R} \frac{k_P}{\gamma_M}(\mu_{SP} - \mu_{YP}),$$

in which n_R is the number of repressor molecules, while K_R is the dissociation constant of the repressor–promoter complex.

8.7 Summary

Here we employed all previous techniques and results to analyze different models for gene expression. We started with very simple toy models, but soon tackled more realistic and more complex models. By progressively increasing the complexity of the models we could understand the consequence of each newly incorporated feature, and learn how to deal with the resulting mathematical complexity. After finishing this chapter the reader is expected to have a more than superficial understanding of the dynamics and thermodynamics of gene expression. But also, he/she is expected to be able to use the results in this chapter as an example of how to use the concepts and results introduced in previous chapter to study all kinds of biochemical systems.

Chapter 9
Ion Channel Dynamics and Ion Transport Across Membranes

Abstract In previous chapters we have employed the formalism and techniques introduced in the book to study different biological systems by conceptualizing them as chemical reactions. In all cases this conceptualization was more or less evident. However, the formalism is more versatile as it can be applied to systems that apparently have nothing to do with chemical reactions. In the present chapter we tackle a few of those systems, all of which are related to diffusion. Not only we exemplify in this chapter how to employ the formalism here introduced to study systems with no obvious connection with chemical reactions, but we also derive some classical results regarding ion transport across membranes. As in previous chapters we start with a very simple example, and gradually make it more complex to end with a more realistic model.

9.1 One Molecule Flipping Between Two Compartment Model

Consider two adjacent compartments I and E, that are connected through a small orifice of area a (let us call it the channel), and whose volumes are respectively V_I and V_E—see Fig. 9.1. Take a single molecule following a Brownian movement across both compartments. In the long run, this molecule shall visit all the locations in both compartments with equal probability (Berg 1993). Therefore, the probability of finding the molecule in compartment I (E) a long time after the experiment started is proportional to V_I (V_E), regardless of the molecule initial position. In other words,

$$\overline{P}_I = \frac{V_I}{V_I + V_E}, \quad \text{and} \quad \overline{P}_E = \frac{V_E}{V_I + V_E}. \tag{9.1}$$

M. Santillán, *Chemical Kinetics, Stochastic Processes, and Irreversible Thermodynamics*, Lecture Notes on Mathematical Modelling in the Life Sciences,
DOI 10.1007/978-3-319-06689-9_9,

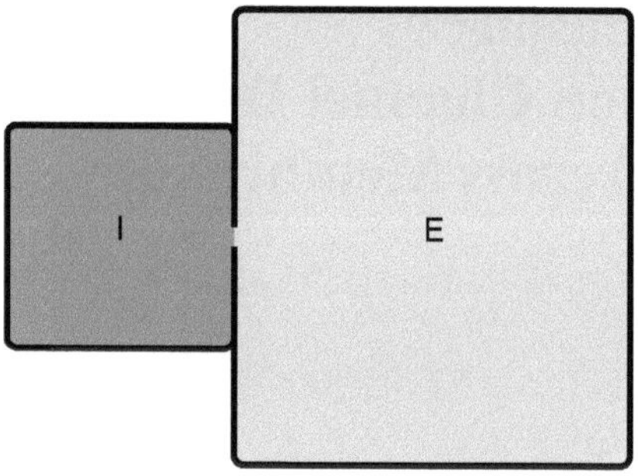

Fig. 9.1 Two reservoirs of different volumes connected through a small orifice

Another way of tackling this problem is to regard all the molecule positions within compartment I as single lumped state, and to do the same for compartment E. With this, the molecule can be seen as randomly flipping back and forth between states I and E. Assume that the transition probabilities from I to E (κ_{IE}), and from E to I (κ_{EI}), are constant. Then, the master equation governing the molecule stochastic dynamics is

$$\frac{dP_I(t)}{dt} = \kappa_{EI}(1 - P_I(t)) - \kappa_{IE} P_I(t), \tag{9.2}$$

with $P_I(t)$ the probability that the molecule is in state I at time t. The probability that the molecule is in state E at time t is simply given by $P_E(t) = 1 - P_I(t)$. Note that the present system is identical to the one studied in Chap. 3 and so the solution to Eq. (9.2) is

$$P_I(t) = \frac{\kappa_{EI}}{\kappa_{IE} + \kappa_{EI}} + \left(P_I^O - \frac{\kappa_{EI}}{\kappa_{IE} + \kappa_{EI}}\right) e^{-(\kappa_{IE}+\kappa_{EI})t}, \tag{9.3}$$

where P_I^O is the initial $P_I(t)$ value. Moreover, as $t \to \infty$, $P_I(t)$ converges to

$$\overline{P}_I(t) = \frac{\kappa_{EI}}{\kappa_{IE} + \kappa_{EI}}, \tag{9.4}$$

while $P_E(t)$ converges to the following stationary value

$$\overline{P}_E(t) = \frac{\kappa_{IE}}{\kappa_{IE} + \kappa_{EI}}. \tag{9.5}$$

Consider the transition rates κ_{IE} and κ_{EI}. Intuitively, one expects that both rates are proportional to the channel area. On the other hand, it seems reasonable to assume that the larger a container is, the longer the molecule takes to scape from it. A very simple way to account for these two suppositions into account is to take

$$\kappa_{IE} = \frac{\alpha}{V_I}, \quad \text{and} \quad \kappa_{EI} = \frac{\alpha}{V_E}, \tag{9.6}$$

with α directly proportional to the channel area. Interestingly, if we substitute Eq. (9.6) into Eqs. (9.4) and (9.5), we recover the results in Eq. (9.1), thus confirming the equivalence of both approaches. Equation (9.6) further suggests to interpret α as the channel conductivity, which is a property of the channel itself and not of the reservoirs it connects.

9.2 *N* Molecules Flipping Between Two Compartments

Consider now a constant number N of molecules, identical to the one in the previous section, which switch independently between compartments I and E. Given molecule independence, the probability of finding n_I molecules in compartment I at time t obeys the following master equation

$$\begin{aligned}\frac{dP(n_I,t)}{dt} &= \frac{\alpha}{V_E}(n_E+1)P(n_I-1,t) - \frac{\alpha}{V_E}n_E P(n_I,t)\\ &+ \frac{\alpha}{V_I}(n_I+1)P(n_I+1,t) - \frac{\alpha}{V_I}n_I P(n_I,t),\end{aligned} \tag{9.7}$$

in which $n_E = N - n_I$ is the molecule count of compartment E. From the results in Chap. 3, the solution of Eq. (9.7) is

$$P(n_I,t) = \frac{N!}{n_I!(N-n_I)!}P_I(t)^{n_I}(1-P_I(t))^{1-n_I}, \tag{9.8}$$

with $P_I(t)$ as given by Eq. (9.3). From this, the probability of finding n_E molecules in E can be computed as

$$P(n_E,t) = P(n_I = N-n_E,t) = \frac{N!}{n_E!(N-n_E)!}P_E(t)^{n_E}(1-P_E(t))^{1-n_E}, \tag{9.9}$$

where $P_E(t) = 1 - P_I(t)$. Notice that $P(n_I,t)$ and $P(n_E,t)$ are complimentary binomial probability distributions.

The mean molecule count in compartment I can be calculated from the corresponding probability distribution as $N_I(t) = \sum_{n_I} n_I P(n_I,t)$. By differentiating this expression and substituting Eq. (9.7) we obtain that

$$\frac{dN_I(t)}{dt} = \alpha\frac{N_E}{V_E} - \alpha\frac{N_I}{V_I}, \tag{9.10}$$

in which $N_E(t) = N - N_I(t)$ is the average number of molecules in compartment E. It is straightforward to prove from the above results that $N_E(t)$ obeys the following differential equation:

$$\frac{dN_E(t)}{dt} = -\frac{dN_I(t)}{dt} = \alpha\frac{N_I}{V_I} - \alpha\frac{N_E}{V_E}. \tag{9.11}$$

From Eqs. (9.10) and (9.11), the average net molecule current from compartment E to compartment I (i.e. the average number of molecules crossing from E to I per unit time) results to be

$$J_{EI}(t) = \alpha(c_E(t) - c_I(t)), \tag{9.12}$$

where $c_I = N_I/V_I$ and $c_E = N_E/V_E$ respectively denote the molecule concentrations in compartments I and E. A quick look at Eq. (9.12) reveals that we have recovered a special case of the well known Fick's law of diffusion in one dimension (Fick 1855): the molecule flow across the channel is directly proportional to the molecule concentration difference. Finally, from Eq. (9.12), the differential Eqs. (9.10) and (9.11) can be rewritten in terms of the flow J_{EI} as $dN_I(t)/dt = -dN_E(t)/dt = J_{EI}(t)$

The readers are encouraged to demonstrate that the solutions to Eqs. (9.10) and (9.11) are:

$$N_I(t) = \overline{N}_I + \left(N_I^O - \overline{N}_I\right)e^{-(\kappa_{IE}+\kappa_{EI})t}, \tag{9.13}$$

$$N_E(t) = \overline{N}_E + \left(N_E^O - \overline{N}_E\right)e^{-(\kappa_{IE}+\kappa_{EI})t}, \tag{9.14}$$

in which

$$\overline{N}_I = N\frac{\kappa_{EI}}{\kappa_{IE}+\kappa_{EI}} = \frac{V_I}{V_I+V_E}, \tag{9.15}$$

$$\overline{N}_E = N\frac{\kappa_{IE}}{\kappa_{IE}+\kappa_{EI}} = \frac{V_E}{V_I+V_E}, \tag{9.16}$$

are the $N_I(t)$ and $N_E(t)$ stationary values, while N_I^O and N_E^O denote the corresponding initial conditions. Interestingly, the same results can be obtained by directly computing $N_I(t)$ and $N_E(t)$ from Eqs. (9.8) and (9.9).

We see from Eqs. (9.15) and (9.16) that

$$\frac{\overline{N}_I}{V_I} = \frac{\overline{N}_E}{V_E}. \tag{9.17}$$

That is, the stationary state is characterized by equal molecule concentrations on both compartments.

To give a thermodynamic interpretation to the result in Eq. (9.17), recall the expression for the chemical potential derived in Chap. 2—Eq. (2.11):

$$\mu = \mu^O + k_B T \ln\frac{c}{c^O}, \tag{9.18}$$

in which c is the concentration of the chemical species of interest, c^O is an arbitrary standard concentration, and μ^O is the energy of a single molecule in solution at concentration c^O. Since the same molecule type is found in compartments I and E, we expect that $\mu_I^O = \mu_E^O = \mu^O$. Then, having equal concentrations in both compartments ($c_I = c_E$) implies that the chemical potentials are the same:

$$\mu_I = \mu_E.$$

In other words, the stationary state is achieved when the chemical potential in both compartments is leveled. This further suggests that the stationary state in this case is concomitant with chemical equilibrium. To corroborate this last assertion consider the Gibbs free energy rate of change—Eq. (3.29):

$$\frac{dG(t)}{dt} = \mu_I(t)\frac{dN_I(t)}{dt} + \mu_E(t)\frac{dN_E(t)}{dt}.$$

Upon substitution of Eqs. (9.10) and (9.11) the above equation becomes

$$\begin{aligned} \frac{dG(t)}{dt} &= \alpha(c_E(t) - c_I(t))(\mu_I(t) - \mu_E(t)) \\ &= -k_B T\alpha(c_E(t) - c_I(t)) \ln\frac{c_E(t)}{c_I(t)}. \end{aligned} \tag{9.19}$$

In the derivation of the last equality we made use of Eq. (9.18). If we consider now the following well-known result:

$$(x - y)\ln(x/y) \begin{cases} < 0, \text{ if } & x, y > 0 \text{ and } x \neq y, \\ = 0, \text{ if } & x, y > 0 \text{ and} x = y \end{cases}$$

it follows from Eq. (9.19) that

$$\frac{dG}{dt} \leq 0 \tag{9.20}$$

when $c_I \neq c_E$, and that $dG/dt = 0$ when $c_I = c_E$. This means that whenever there is a molecule concentration unbalance between both compartments, the system free energy tends to decrease until reaching its minimum value at the stationary state (when both concentrations are equal). Since we know that, under conditions of constant pressure, temperature, and molecule count, the thermodynamic equilibrium state is that which minimizes Gibbs free energy, the above result confirms that the system stationary state complies with chemical equilibrium.

9.3 Constant Concentration Gradient Across the Wall Separating Both Compartments

We have seen that when there is a constant number of molecules randomly flipping back and forth between two compartments, the stationary state is reached when the molecule concentrations in both compartments are leveled. Furthermore, this situation corresponds to thermodynamic equilibrium because the net average molecule flux between compartments vanishes, and because the system free energy reaches a constant minimum value.

In this section we analyze a different situation: that in which the molecule concentrations in both compartments are kept constant along time. Let c_I and c_E respectively denote the molecule concentrations in compartments I and E, and assume without loss of generality that $c_E > c_I$. Hence, according to Eq. (9.12), there exists a constant net average molecule current from compartment E into compartment I:

$$J_{EI} = \alpha(c_E - c_I), \tag{9.21}$$

which emerges as a result of the constancy of c_I and c_E. To compensate this constant molecule flow, new molecules need to be continually added to the higher concentration compartment, and removed from the lower concentration one.

From a thermodynamic perspective, each time a molecule goes from the high to the low concentration compartments, its free energy decreases (recall that chemical potential can be understood as free energy per molecule in a given compartment). The free energy reduction per transported molecule is

$$\Delta\mu = k_B T \ln \frac{c_E}{c_I}. \tag{9.22}$$

This energy reduction is ultimately dissipated as heat. If we consider the molecule current from E to I by Eq. (9.21), we can deduce the following expression for the heat dissipation rate:

$$\phi = k_B T \alpha(c_E - c_I) \ln \frac{c_E}{c_I} > 0. \tag{9.23}$$

In spite of heat being continuously dissipated, the system is in a stationary state both dynamically and thermodynamically. From a thermodynamical point of view, the stationary state is maintained because by energy constantly pumped into the system by adding new molecules into the high concentration compartment and removing them from the low concentration one. Finally, due to the constant molecule flow and heat dissipation, the achieved stationary state is not compatible with chemical equilibrium.

9.4 Two Compartments Connected by a Channel with a Driving Force

Consider once more the system pictured in Fig. 9.1. Further assume now that the channel connecting both compartments has length l, and that a constant conservative force of magnitude F, pointing to direction E to I, is exerted upon all molecules within the channel. Given the conservativeness of F, each time a molecule crosses the channel from E to I, its potential energy decreases by an amount $V = Fl$, regardless of its trajectory. Similarly, each time a molecule crosses the channel from I to E, its potential energy is increased by the same amount. If we choose the zeroth level of potential energy at the channel midpoint, the potential energy per molecule in the I and E compartments is $-V/2$ and $V/2$, respectively. On the other hand, recall that in the absence of the extra driving force F, the energy per molecule is μ_O in both compartments. Hence, when the force F is included, the total energy of each molecule in compartment I is $\mu^O - V/2$, while every molecule in compartment E has energy $\mu^O + V/2$. From these considerations, the chemical potentials in both compartments result to be:

$$\mu_I(t) = \mu^O - \frac{V}{2} + k_B T \ln N_I(t), \tag{9.24}$$

$$\mu_E(t) = \mu^O + \frac{V}{2} + k_B T \ln N_E(t). \tag{9.25}$$

Intuitively, the driving force F is expected to increase molecule concentration in compartment I, and to decrease it in E. However, as soon as it appears, this concentration unbalance makes molecules flow in the opposite direction (I to E). Hence, a stationary situation should be reached in which the molecule flows caused by force F and by the concentration unbalance cancel each other. Moreover, given that once the stationary state is reached, no net molecule flow exists in either direction, the steady state should be compatible with chemical equilibrium and so the chemical potentials on both compartments must be equal. This further implies that

$$\overline{c}_I = \overline{c}_E e^{V/k_B T}. \tag{9.26}$$

This equation can be more clearly appreciated if we rewrite it as

$$k_B T(\ln \overline{c}_I - \ln \overline{c}_E) = V.$$

We can see now how the potential energy difference V is compensated by an *entropic* concentration unbalance.

The same problem can be tackled from a dynamical, rather than a thermodynamical, perspective. To this end recall that, in the absence of force F, the propensities with which one molecule switches from I to E, and from E to I, are those given by Eq. (9.6). On the other hand, because of force F the energy of each molecule in E is increased by an additive term $V/2$, while the energy of every molecule in I decreases by the same amount. This in turn increases the propensity of E to I transitions because the energy barrier a molecule needs to surpass is now lower, and decreases the propensity of I to E transitions because the corresponding energy barrier is now larger. Given the relation between energy barriers and propensities—Eqs. (3.29) and (3.30), the above discussed facts imply that the propensities in the presence of force F are now given by

$$\kappa'_{IE} = \frac{\alpha}{V_I} e^{-V/2k_B T}, \quad \text{and} \quad \kappa'_{EI} = \frac{\alpha}{V_E} e^{V/2k_B T}. \tag{9.27}$$

By repeating the analysis leading to Eq. (9.11) with the propensities in Eq. (9.27) we obtain the following differential equations for the dynamics of the average molecular counts in I and E:

$$\frac{dN_E(t)}{dt} = -\frac{dN_I(t)}{dt} = \kappa'_{IE} N_I - \kappa'_{EI} N_E. \tag{9.28}$$

The solutions of these differential equations are

$$N_I(t) = \overline{N}_I + \left(N_I^O - \overline{N}_I\right) e^{-(\kappa'_{IE}+\kappa'_{EI})t}, \tag{9.29}$$

$$N_E(t) = \overline{N}_E + \left(N_E^O - \overline{N}_E\right) e^{-(\kappa'_{IE}+\kappa'_{EI})t}, \tag{9.30}$$

in which

$$\overline{N}_I = N \frac{\kappa'_{EI}}{\kappa'_{IE} + \kappa'_{EI}}, \tag{9.31}$$

$$\overline{N}_E = N \frac{\kappa'_{IE}}{\kappa'_{IE} + \kappa'_{EI}}, \tag{9.32}$$

are the $N_I(t)$ and $N_E(t)$ stationary values, while N_I^O and N_E^O represent the corresponding initial conditions. By simple substitution it is straightforward to prove that Eqs (9.31) and (9.32) are in complete agreement with Eq. (9.26), confirming that the stationary state complies with thermodynamic equilibrium.

9.5 A Randomly Gating Channel

Consider again the picture in Fig. 9.1. Assume without loss of generality that the molecule concentrations on both compartments are constant and obey $c_E > c_I$. Finally, suppose that the channel can be either open or closed, and that it randomly

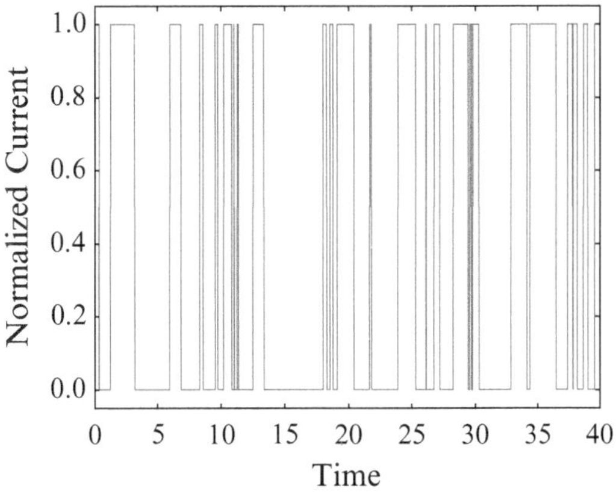

Fig. 9.2 Gillespie simulation of the current through a randomly gating channel with transition propensities $k_{co} = 1$ and $k_{oc} = 2$

flips between its open and closed states; the propensities of the closed-to-open and the open-to-closed transitions respectively being k_{co} and k_{oc}. Under these considerations, the channel corresponds to the system studied in Chap. 3. Hence, the stationary probabilities of finding a channel in the open and closed states are respectively given by:

$$P_o = \frac{k_{co}}{k_{co} + k_{oc}}, \tag{9.33}$$

$$P_c = \frac{k_{oc}}{k_{co} + k_{oc}}. \tag{9.34}$$

Furthermore, the times the channel remains open or closed are both stochastic variables which obey exponential distributions with average values

$$\tau_o = \frac{1}{k_{oc}}, \tag{9.35}$$

$$\tau_c = \frac{1}{k_{co}}. \tag{9.36}$$

Finally, each time the channel opens, a constant current

$$J_{EI} = \alpha(c_E - c_I)$$

in the E to I direction appears, while when the channel is closed the current is zero. Therefore, a way to experimentally investigate the channel dynamics would be to record the current. A Gillespie simulation of the current in one of such experiments is shown in Fig. 9.2.

9.6 N Randomly Gating Channels

Let us now consider the dynamics of N channels identical to the one studied in the previous section. Let n_o and $n_c = N - n_o$ respectively denote the number of open and close channels, and $P(n_o, t)$ the probability of having n_o open channels at time t. Following Chap. 3, the master equation governing the dynamics of $P(n_o, t)$ is

$$\frac{dP(n_o,t)}{dt} = k_{oc}(n_o+1)P(n_o+1,t) - k_{oc}n_oP(n_o,t) + k_{co}(N-n_o+1)P(n_o-1,t) - k_{co}(N-n_o)P(n_o,t). \tag{9.37}$$

From the results in Chap. 3, the stationary solution to this equation is the following binomial distribution:

$$\overline{P}(n_o) = \frac{N!}{n!(N-n_o)!}P_o^n(1-P_o)^{N-n_o}, \tag{9.38}$$

with P_o as given by Eq. (9.33). The average number of open channels resulting from this distribution is NP_o, while the corresponding standard deviation is $\sqrt{NP_o(1-P_o)}$. Hence, the coefficient of variation is

$$CV = \sqrt{\frac{1-P_o}{P_o}\frac{1}{\sqrt{N}}}.$$

We see that the coefficient of variation is inversely proportional to $\sqrt{N}$. Thus, for very large N, the average n_o value is a good descriptor of the system dynamics.

By definition, the average n_o value is given by

$$N_o(t) = \sum_{n_o} n_o P_o(n_o, t).$$

Then,

$$\frac{dN_o(t)}{dt} = \sum_{n_o} n_o \frac{dP(n_o,t)}{dt}.$$

Substitution of Eq. (9.37) into the above equation leads to the following differential equation for the dynamics of $N_o(t)$:

$$\frac{dN_o(t)}{dt} = k_{co}(N - N_o(t)) - k_{oc}N_o(t). \tag{9.39}$$

The solution to the above equation is:

$$N_o(t) = \frac{k_{co}}{k_{oc} + k_{co}} N + \left(N_o(0) - \frac{k_{co}}{k_{oc} + k_{co}} N \right) e^{-(k_{oc}+k_{co})t}. \tag{9.40}$$

We see from Eq. (9.40) that no matter what the initial condition is, $N_o(t)$ exponentially converges to the stationary state

$$\overline{N}_o = \frac{k_{co}}{k_{oc} + k_{co}} N,$$

with rate $k_{oc} + k_{co}$.

Recall that the current from E to I through each open channel is $\alpha(c_E - c_I)$, with c_E and c_I the corresponding molecule concentrations. Hence, if at any given time the number of open channels is n_o, the total current from E to I is $\alpha n_o(c_E - c_I)$. Since the channels are randomly opening and closing, the number of open ones is a random variable whose value fluctuates in time. However, when the number of channels is very large, the average number of open channels becomes a good descriptor of the system dynamics because the size of fluctuations is negligible as compared with the mean value. In that scenario, the total current will slightly fluctuate around the following average value:

$$J_{EI}(t) = \frac{k_{co}}{k_{oc} + k_{co}} J_{EI}^{\max} + \left(J_{EI}(0) - \frac{k_{co}}{k_{oc} + k_{co}} J_{EI}^{\max} \right) e^{-(k_{oc}+k_{co})t}, \tag{9.41}$$

where

$$J_{EI}^{\max} = \alpha N(c_E - c_I)$$

is the maximal possible current (which takes place when all channels are open). The product αN can be interpreted as the maximal system conductance, obtained as the product of each channel conductivity times the channel count. Equation (9.41) further implies that the average current is modulated by the probability that a given channel is open at a given time

$$P_o = \frac{k_{co}}{k_{oc} + k_{co}},$$

which in turn is a function of the transition rates. Therefore, if one could manipulate the transition rates between the open and close states it would be possible to increase or decrease the current J_{EI}. We shall come back to this point in the next section.

To close this section, let us briefly analyze the problem from the perspective of thermodynamics. The whole system can be seen as composed of two subsystems. On the one hand we have the channels, which randomly gate between the open and closes states, and on the other hand we have the net flow of molecules from the high concentration and into the low concentration compartments. The first subsystem complies with thermodynamic equilibrium in the stationary state (the chemical potential of open and closed channels is identical), and so there is no heat dissipation associated with channel gating. Regarding the flow of molecules, recall that the system free energy is

$$G = \mu_E N_E + \mu_I N_I,$$

where N_E and N_I are the molecule counts in E and I, while μ_E and μ_I are the corresponding chemical potentials. Then, each time a molecule goes from E to I, the system free energy changes by the following amount

$$\Delta G = -(\mu_E - \mu_I).$$

The first law of thermodynamics implies that this energy decrease is dissipated as heat into the environment. Finally, given that the current J_{EI} measures the net number of molecules shifting from E to I per unit time, the heat dissipation rate associated with this phenomenon is

$$\phi(t) = -J_{EI}(t)\Delta G.$$

In particular, the stationary heat dissipation rate is given by

$$\overline{\phi} = \alpha N \frac{k_{co}}{k_{oc} + k_{co}} (c_E - c_I)(\mu_E - \mu_I). \tag{9.42}$$

If we further take into consideration that $\mu_E = \mu^O + k_B T \ln N_E$ and $\mu_I = \mu^O + k_B T \ln N_I$, Eq. (9.42) can be rewritten as

$$\overline{\phi} = k_B T \alpha N \frac{k_{co}}{k_{oc} + k_{co}} (c_E - c_I) \ln \frac{c_E}{c_I} \geq 0. \tag{9.43}$$

Interestingly, the heat dissipation rate is always positive when $c_E \neq c_I$, regardless of which concentration is larger. Furthermore, in order to maintain the stationary state, new molecules need to be continuously added to the high concentration compartment and removed from the low concentration one. By doing this, one also compensates for the dissipated energy because high energy molecules are constantly incorporated into the system and low energy ones are removed.

9.7 Ion Channel Regulation

We have seen in the previous sections how a totally random process like the gating of an ion channel between the open and the closed states can give rise to a deterministic phenomenon at the macroscopical level: the flow of molecules from E to I, when

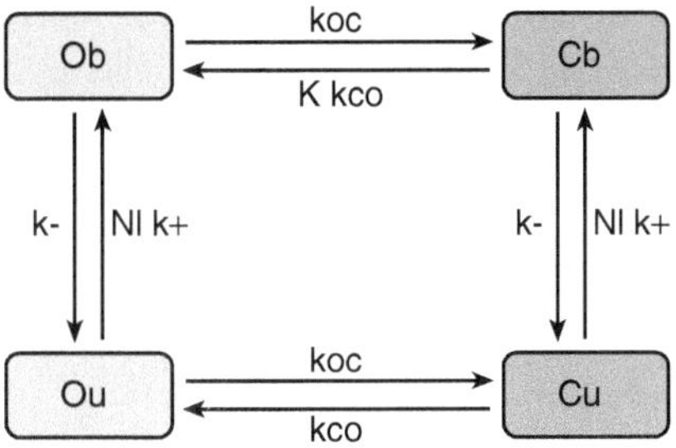

Fig. 9.3 Schematic representation of the available states, and the transitions between them, for a channel that can be either open (O) or closed (C), and that in each state can be either unbound (u) or bound (b) by a regulatory molecule. N_l denotes the regulatory molecule count

there are numerous channels connecting the two compartments. Interestingly, the magnitude of the molecule flow depends not only on the total number of channels, on their conductivity, and on the molecule gradient across the channels, but also on the channel transition rates between the open and closed states. Hence, in order to control the flow of molecules through the channels one could modify either one of the above-mentioned characteristics. Let us analyze each one of them. The conductivity is a property of the channels themselves and so it is not possible to change it without modifying the channel chemical nature. Since modifying the number of channels involves producing or degrading membrane proteins, this is not something that could be done in the time scale of seconds or less, as it is sometimes required. Finally, in some occasions changing the concentrations of the transported molecule is not an option. Thus, the only remaining characteristics are the transition rates between the channel open and closed states. Is it possible to modify them? As a matter of fact it is and nature has found several ways to do it. In what follows we shall study a simple instance of one of the most common mechanisms in that respect: the binding of small molecules to the channels to modify their open probability.

Consider a channel that can be either open (O) or closed (C), and that in each state it can be either unbound (u) or bound (b) by a regulatory molecule. The four available state for this channel, as well as the transitions between them, are pictured in Fig. 9.3. The four states accessible to the channel are: open and bound by the regulatory molecule (Ob), open and unbound by the regulatory molecule (Ou), closed and bound by the regulatory molecule (Cb), and closed and unbound by the regulatory molecule (Cu). Assume that the propensity of the transition from bound to unbound is proportional to the number of regulatory molecules (N_l), but independent of whether the channel is open or closed; let k^+ be the proportionality constant. Assume also that the propensity of the bound to unbound transition (k^-) is independent on the channel (open or closed) state. Further suppose that the propensity of the open to closed transition k_{oc} is the same for bound and unbound channels. Finally, if k_{co} represents the propensity of the closed to open transition for an unbound channel, assume that the propensity for the corresponding transition in a bound channel is Kk_{co}, with $K > 1$. That is, the binding of the regulatory molecule facilitates the transition of the channel to the open state. With the above considerations, the master equations governing the dynamics of the probabilities of finding the channel in each of its four available states are:

$$\frac{dP(Ob,t)}{dt} = Kk_{co}P(Cb,t) - k_{oc}P(Ob,t) + k^{+}N_l P(Ou,t) - k^{-}P(Ob,t), \tag{9.44}$$

$$\frac{dP(Cb,t)}{dt} = k_{oc}P(Ob,t) - Kk_{co}P(Cb,t) + k^{+}N_l P(Cu,t) - k^{-}P(Cb,t), \tag{9.45}$$

$$\frac{dP(Ou,t)}{dt} = k_{co}P(Cu,t) - k_{oc}P(Ou,t) + k^{-}P(Ob,t) - k^{+}N_l P(Ou,t), \tag{9.46}$$

$$\frac{dP(Cu,t)}{dt} = k_{oc}P(Ou,t) - k_{co}P(Cu,t) + k^{-}P(Cb,t) - k^{+}N_l P(Cu,t). \tag{9.47}$$

Interestingly, the above is a redundant system of differential equations because

$$P(Cu,t) = 1 - P(Ob,t) - P(Cb,t) - P(Ou,t).$$

Therefore

$$\frac{dP(Cu,t)}{dt} = -\frac{dP(Ob,t)}{dt} - \frac{dP(Cb,t)}{dt} - \frac{dP(Ou,t)}{dt}.$$

As a result of the above, we have to deal with a 3-dimensional system, rather than with a 4-dimensional one. Nonetheless, this system is still too complex to be analytically studied. One way to simplify the system is to suppose that the unbound-to-bound transitions are much faster than those between the open and closed states, and use the quasi-stationary approximation introduced in Chap. 5. Let us define the probabilities that the channel is open and closed (regardless of their being bound or unbound by the regulatory molecule) as follows:

$$P(O,t) = P(Ob,t) + P(Ou,t), \quad P(C,t) = P(Cb,t) + P(Cu,t). \tag{9.48}$$

According to the methodology introduced in Chap. 5, after taking advantage of the separation of time scales to reduce the complexity of a system, the equation governing the dynamics of $P(O,t)$ results to be

$$\frac{dP(O,t)}{dt} = \kappa_{co}P(C,t) - \kappa_{oc}P(O,t), \tag{9.49}$$

In which

$$\kappa_{oc} = k_{oc}(\overline{P}(Ob|O) + \overline{P}(Ou|O)) = k_{oc} \tag{9.50}$$

and

$$\begin{aligned}\kappa_{co} &= k_{co}(K\overline{P}(Cb|C) + \overline{P}(Cu|C)) \\ &= k_{co}\left(K\frac{k^+N_l}{k^+N_l + k^-} + \frac{k^-}{k^+N_l + k^-}\right).\end{aligned} \tag{9.51}$$

We can see from the above equation that κ_{co} is a growing function of the regulatory-molecule count. The lower and upper limits, reached at $N_l = 0$ and $\lim N_l \to \infty$, are:

$$k_{co} \leq \kappa_{co} \leq Kk_{co}. \tag{9.52}$$

Regarding the probability that the channel is open at time t, it is given by

$$P(C,t) = 1 - P(O,t). \tag{9.53}$$

Thus

$$\frac{dP(C,t)}{dt} = -\frac{dP(O,t)}{dt}. \tag{9.54}$$

From Eq. (9.49), the stationary open probability distribution is given by

$$\overline{P}(O) = \frac{\kappa_{co}}{\kappa_{co} + \kappa_{oc}}. \tag{9.55}$$

Notice that, through κ_{co}, $\overline{P}(O)$ is a growing function of N_l. Moreover, the lower and upper bounds for $\overline{P}(O)$ are:

$$\frac{k_{co}}{k_{co} + k_{oc}} \leq \overline{P}(O) \leq \frac{k_{co}K}{k_{co}K + k_{oc}} \tag{9.56}$$

If we consider N identical channels and we repeat the procedure leading to Eq. (9.39), we obtain the following expression for the differential equation governing the dynamics of the average number of open channels ($N_O(t)$):

$$\frac{dN_O(t)}{dt} = \kappa_{co}(N - N_O(t)) - \kappa_{oc}N_O(t). \tag{9.57}$$

It is straightforward to prove from the above equation that the N_O stationary value is given by

$$\overline{N}_O = N\overline{P}(O) = N\frac{\kappa_{co}}{\kappa_{co} + \kappa_{oc}}. \tag{9.58}$$

Since κ_{co} is a growing function of the ligand count N_l, the average number of open channels can be increased by augmenting N_l. Furthermore, from (9.58), the lower and upper limits for $\overline{N}_O$ (reached at $N_l = 0$ and $\lim N_l \to \infty$, respectively) are

$$N\frac{k_{co}}{k_{co} + k_{oc}} \leq \overline{N}_O \leq N\frac{k_{co}K}{k_{co}K + k_{oc}} \tag{9.59}$$

Finally, since $\overline{N}_O$ determines the molecule flow from E to I, it follows that the current J_{EI} can be controlled by changing the number of regulatory molecules N_l. As a matter of fact, a larger number of regulatory molecules imply a larger current J_{EI}, which has the following lower and upper bounds:

$$\alpha N(c_E - c_I)\frac{k_{co}}{k_{co} + k_{oc}} \leq J \leq \alpha N(c_E - c_I)\frac{k_{co}K}{k_{co}K + k_{oc}} \tag{9.60}$$

9.8 Summary

The results and concepts introduced in the first chapters of this book can be directly applied to the study of any biological system that can be conceptualized as a system of chemical reactions. This is more or less straightforward in many cases, as illustrated in Chap. 8. However, there are some other cases where the connection is not evident, yet the approaches we have been using here can be extremely helpful. In this chapter we illustrate that by carefully studying ion transport across membranes and ion channel regulation. Although the chapter was not meant as a treatise on electro-physiology, it exemplifies how by combining chemical kinetics, irreversible thermodynamics, and stochastic processes, one can recover some of the most important results in that science.

References

Beard D, Qian H (2008) Chemical biophysics: quantitative analysis of cellular systems. Cambridge Texts in Biomedical Engineering. Cambridge University Press, Cambridge
Ben-Naim A (2007) Entropy demystified: the second law reduced to plain common sense. World Scientific, New Jersey
Berg H (1993) Random walks in biology. Princeton paperbacks, Princeton University Press, Princeton
Cannon WB (1929) Organization for physiological homeostasis. Physiol Rev 9(3):399–431
E W, Vanden-Eijnden E (2010) Transition-path theory and path-finding algorithms for the study of rare events. Ann Rev Phys Chem 61(1):391–420
Evans M, Hastings N, Peacock J (2000) Statistical distributions. Wiley Series in Probability and Statistics. Wiley, New York
Fick A (1855) Ueber diffusion. Ann Phys 170(1):59–86
Gillespie DT (1977) Exact stochastic simulation of coupled chemical reactions. J Phys Chem 81(25):2340–2361
Golding I, Paulsson J, Zawilski SM, Cox EC (2005) Real-time kinetics of gene activity in individual bacteria. Cell 123:1025–1036
de Groot S, Mazur P (2013) Non-equilibrium thermodynamics. Dover Publications, Mineola
Houston P (2001) Chemical kinetics and reaction dynamics. McGraw-Hill, New York
Jaynes ET (2003) Probability theory: the logic of science. Cambridge University Press, Cambridge
Lehninger A, Nelson D, Cox M (2005) Lehninger principles of biochemistry. W. H. Freeman, New York
McClure WR (1985) Mechanism and control of transcription initiation in prokaryotes. Ann Rev Biochem 54:171–204
Planck M (1945) Treatise on thermodynamics. Dover Books on Physics Series. Dover Publications, Mineola
Risken H (1996) The Fokker-Planck equation: methods of solution and applications. Springer, Berlin
Ross SM (1983) Stochastic processes. Wiley Series in Probability and Mathematical Statistics. Wiley, New York
Santillán M (2008) On the use of the hill functions in mathematical models of gene regulatory networks. Math Mod Nat Phenom 3:85–97
Santillán M, Qian H (2011) Irreversible thermodynamics in multiscale stochastic dynamical systems. Phys Rev E 83:041,130
Shahrezaei V, Swain PS (2008) Analytical distributions for stochastic gene expression. Proc Natl Acad Sci 105(45):17,256–17,261

M. Santillán, *Chemical Kinetics, Stochastic Processes, and Irreversible Thermodynamics*, Lecture Notes on Mathematical Modelling in the Life Sciences,
DOI 10.1007/978-3-319-06689-9,

Strogatz S (1994) Nonlinear dynamics and chaos: with applications to physics, biology, chemistry and engineering. Studies in Nonlinearity Series. Perseus Books Publishing, New York

Van Kampen N (1992) Stochastic processes in physics and chemistry. North-Holland Personal Library. Elsevier Science, Amsterdam

Zeron ES, Santillán M (2010) Distributions for negative-feedback-regulated stochastic gene expression: Dimension reduction and numerical solution of the chemical master equation. J Theor Biol 264(2):377–385

Zill DG (2008) A first course in differential equations with modeling applications. Brooks/Cole, Cengage Learning, Belmont

Index

M. Santillán, *Chemical Kinetics, Stochastic Processes, and Irreversible Thermodynamics*, Lecture Notes on Mathematical Modelling in the Life Sciences, DOI 10.1007/978-3-319-06689-9,